Seed Science and Technology

Questions & Answers

Seed Science and Technology

Questions & Answers

By :

K. RAJA

K. SIVASUBRAMANIAM

R. GEETHA

Department of Seed Science & Technology
Agricultural College & Research Institute, Madurai-625 104

SATISH SERIAL PUBLISHING HOUSE

403, Express Tower, Commercial Complex, Azadpur, Delhi-110033 (India)
Phone : 011-27672852, Fax : 91-11-27672046
E-mail :info@satishserial.com, hkjain1975@yahoo.com
Website : www.satishserial.com

Published by :

SATISH SERIAL PUBLISHING HOUSE
403, Express Tower, Commercial Complex, Azadpur, Delhi-110033 (INDIA)
Phone : 011-27672852 Fax : 91-11-27672046
E-mail : info@satishserial.com, hkjain1975@yahoo.com

ISBN 978-93-81226-21-6

Typeset at: Harminder Singh Kharb *for* Laxmi Art Creation
Printed at: Salasar Imaging Systems, Delhi-110035

Preface

Seed is an important input in agriculture and any agriculture begins with the seed. Seed formation involves several sciences beginning from syngamy (fertilization) until it has been sown in the field. Normally the seeds are sown by mankind for cultivation of crops and sometimes it is dispersed autochory that the seed is capable of. Also, germination is decided by the seed itself due to its dormancy or non-dormant nature. Recalcitrancy is another feature in some kinds of seed, which decides the storability of that seed. Generally the orthodox seeds can be stored well when compared to recalcitrant seed.

Many organizations are involved in the production and quality control system of crop seeds throughout the world. The seed testing procedures described by ISTA are being followed by our country. The Seeds Act and Seed Rules are the legislative laws being followed throughout India for the assurance of seed quality.

Therefore in this book, maximum possible seed science related questions have been included followed by their answers. The questions and answers are prepared with single or multiple answers which will be helpful to the scientists and students working in the field of seed science. Reader anticipation has been reduced with answers following questions. This may help in quick check. Also the varietal and hybrid seed production techniques and the related terminologies are highly useful to the scientists and other technical persons engaged in this area. Since this book is prepared with latest information and we are sure that the book will be insightful to the students for updating their knowledge and also those preparing for examinations.

Authors

Contents

Chapter 1

Seed Biology

Q.1 Botanically seed is a ________

Ans. Ripened ovule

Q.2 What is meant by seed in agriculture?

Ans. Seed is any part of organ of a plant which has the capability to regenerate a new plant and can be sown or distributed without much difficulty.

Q.3 'True seed' or 'sexual seed' is meant by ________

Ans. Seed developed from sexual fertilization of male and female gametes.

Q.4 The term 'apomictic seed' refers to ________

Ans. Seed with haploid embryo produced without fertilization.

Q.5 Botanically fruit is termed as a ________

Ans. Ripened ovary.

Q.6 'Naked' seed is meant by _______

Ans. Seed developed from naked ovule

Q.7 What is meant by covered seed?

Ans. Seed developed within the ovary

Q.8 __________ plant produces naked seed.

Ans. Gymnosperm

Q.9 Which plant produces covered seed?

Ans. Angiosperms

Q.10 Smallest seed in the plant kingdom is__________

Ans. Orchid seed

Q.11 Weight of the orchid seed?

Ans. 20.33 μg

Q.12 Largest seed in the world is __________

Ans. Double coconut (*Lodoicea maldivicea*), Coco-de-mer

Q.13 Origin of this largest seed is ________

Ans. Seychelles Island, Indian Ocean

Q.14 Size of the largest seed?

Ans. Size : 30 cm long and 20 cm thick
weight : 16 kg

Q.15 How many years does the largest seed take to mature?

Ans. 7 years

Q.16 How many years does the largest seed takes to germinate?

Ans. 2 years

Q.17 What are all the methods of propagation?

Ans. Sexual and asexual

Q.18 What is meant by asexual propagation?

Ans. Propagation by using vegetative propagules

Q.19 Examples for vegetative propagules?

Ans. Tubers, setts, suckers, bulbs, corms, cuttings, grafts, buddings

Q.20 Specify the crops in which tubers and rhizomes are used as seed?

Ans. Potato, colocasia, yam, turmeric, gladiolus, tuberose, ginger

Q.21 In onion which part is used for vegetative propagation?

Ans. Bulbs

Q.22 Name some crops in which setts are used for propagation?

Ans. Tapioca, sugarcane

Q.23 What is the name of the sugarcane true seed?

Ans. Fluff

Q.24 What is the name of the rose seed?

Ans. Hips

Q.25 What are seeds with unfunctional cotyledons known as?

Ans. Acotylar

Q.26 Mention the crops propagated by suckers.

Ans. Pineapple, chrysanthemum, banana.

Q.27 What is the component present in greengram seed coat for its green colour?

Ans. Chlorophyll pigments

Q.28 What is the colour of the onion seed?

Ans. Black

Q.29 Give examples for monoecious crop plants.

Ans. Maize, cucumber, pumpkin, ribbed gourd, snake gourd, castor, bittergourd

Q.30 What is the name of the monocot seed?

Ans. Caryopsis

Q.31 The name of the rice fruit or seed is called__________.

Ans. Covered spikelet (Caryopsis)

Q.32 Provide an example for covered spikelet?

Ans. Rice

Q.33 Provide and example for free caryopsis?

Ans. Wheat

Q.34 How many seeds are found in monogerm beet?

Ans. One

Q.35 The fruit of the blackgram is called __________

Ans. Pod

Q.36 The sesame fruit is called __________

Ans. Capsule

Q.37 What is the name of the tomato fruit?

Ans. Berry

Q.38 Mention the name of the cruciferous fruits?

Ans. Siliqua

Q.39 The fruit of sunflower is known as__________

Ans. Achene

Q.40 The marigold seed/fruit is called __________

Ans. Achene

Q.41 What is the name of a winged achene ?

Ans. Samara

Q.42 What is the name given to a coriander fruit?

Ans. Schizocarp

Q.43 What is the storage tissue of monocot seeds?

Ans. Endosperm

Q.44 What is the chromosome number of endosperm?

Ans. 3n (Triploid)

Q.45 Endosperm tissue is viable or non-viable?

Ans. Non-viable

Q.46 Which part of an ovule develops into endosperm during seed development?

Ans. Polar nuclei

Q.47 Chromosome number of embryo?

Ans. 2n (Diploid)

Q.48 Miniature of plant in seed is called?

Ans. Embryo

Q.49 'Lomentum' is referred as ____________

Ans. Partition wall / carpallery wall that separates each seed as in leguminaceae.

Q.50 The cap covers the radicle in dicot seed is known as ____

Ans. Rootcap

Q.51 The cap covers the coleorhiza in monocot seed is known as ___

Ans. Root primordia

Q.52 Indicate the essential parts of a embryo.

Ans. Radicle, plumule, hypocotyl in dicot seed embryo.
Coleorhiza, coleoptile, mesocotyl in monocot seed embryo.

Q.53 Which part of the embryo develops into roots?

Ans. Radicle/Coleorhiza

Q.54 Plumule or coleoptile of the embryo develops into ____?

Ans. Shoots

Q.55 In monocot seeds, the cotyledon is otherwise called______?

Ans. Scutellum

Q.56 Black colour layer present in periphery of monocot endosperm is known as __

Ans. Aleurone layer

Q.57 What is meant by aleurone layer?

Ans. It is a protein body in monocot seed.

Q.58 Functions of aleurone layer?

Ans. Release of enzymes.

Q.59 What is the cause of colour in red rice?

Ans. Pigmented aleurone layer

Q.60 What is the name of monocot seeds based on the storage tissue?

Ans. Endospermous

Q.61 Endospermous seeds are otherwise called as __________

Ans. Albuminous seeds

Q.62 Where do you find carpophore?

Ans. Schizocarp of Umbelliferae family

Q.63 What is the name of monocot seed coat?

Ans. Pericarp

Q.64 What is the name of the appendage present in rice seed?

Ans. Awn

Q.65 Awn is an extension of tip of ______ glume.

Ans. Lemma

Q.66 Mention the parts of a dicot seed?

Ans. Seed coat, cotyledon, embryo.

Q.67 What is the protective part of a seed?

Ans. Seed coat

Q.68 Other name of seed coat is ______

Ans. Spermoderm

Q.69 Indicate the other name of seed coat ?

Ans. Testa

Q.70 What are the names of outer and inner seed coats?

Ans. Outer seed coat - testa

Inner seed coat - tegmen

Q.71 The mucilaginous substance present around the papaya seed is ______

Ans. Sarcotesta

Q.72 The chemical deposit in the seedcoat is ______

Ans. Hemi-cellulose

Q.73 What part of the ovary develops into seed coat?

Ans. Integument

Q.74 Storage tissue of dicot seeds are termed as______

Ans. Cotyledon

Q.75 What is the name of the dicot seeds based on storage tissue?

Ans. Cotyledonary seed or non-endospermous seed.

Q.76 Non endospermous seed is also termed as ______

Ans. Exalbuminous seed

Q.77 Name few dicots seeds with endosperm?

Ans. Castor, fenugreek, tomato, chilli

Q.78 Which part of the ovule develops into perisperm?

Ans. Nucellus

Q.79 The rudimentary endosperm in dicot seed is called as________

Ans. Perisperm

Q.80 The unutilized remains of nucellus in dicot seed are termed as__________

Ans. Perisperm

Q.81 What is the storage organ in amaranthus?

Ans. Perisperm

Q.82 Specify some crop seeds that have perisperm?

Ans. Pepper, Beet

Q.83 The storage organ in pepper seed is ______

Ans. Perisperm

Q.84 What is the name of the scar of attachment of dicot seed with mother plant?

Ans. Hilum

Q.85 The hole present in the seed coat near hilum region________

Ans. Micropyle or germpore

Q.86 What is meant by micropyle in a seed?

Ans. It is a hole through which water enters.

Q.87 What is the name of the outgrowth below the hilum?

Ans. Strophiole.

Q.88 In which crop seed the strophiole can be seen?

Ans. Luguminous seeds.

Q.89 What is the name of the longitudinal whitish ridge on the seed?

Ans. Raphe

Q.90 What is meant by fibre in cotton seed?

Ans. Single celled elongated structure

Q.91 Name of the short fibres on the cotton seed?

Ans. Fuzz

Q.92 The small pair of structures present on the proximal end of rice seed is _____

Ans. Glumes

Q.93 Tip of the glumes is known as ______

Ans. Apiculus

Q.94 Name of the two glumes on rice spikelet are __________.

Ans. Lemma and palea

Q.95 The broader glume of the rice spikelet is called ________

Ans. Lemma

Q.96 Specify the crop seeds having tight glumes?

Ans. Italian millet, Kodo millet

Q.97 What is the name of the rice pericarp?

Ans. Bran

Q.98 What is the main chemical component of a monocot endosperm?

Ans. Carbohydrate/Starch

Q.99 The tissue that joins the seed with mother plant is ________

Ans. Funicle

Q.100 Degenerated funicular scar in monocot seed is called ____________

Ans. Dunken layer

Q.101 The colour of the dunken layer in maize and sorghum seeds?

Ans. Black

Q.102 Respiration rate of seed is measured by the instrument called ___________.

Ans. Manometer

Q.103 Whether the synthetic seed is a sexual or asexual seed?

Ans. Asexual seed

Q.104 What is meant by synthetic seed?

Ans. Synthetic seed is a encapsulated somatic embryo

Q.105 Chemicals used for encapsulation of somatic embryo in synthetic seed?

Ans. Sodium alginate, calcium chloride

Q.106 Who discovered synthetic seeds?

Ans. Keith Redenbaugh

Q.107 Name the protein present in grain legume seeds?

Ans. Phaseolin

Q.108 Name the protein molecule present in maize seed?

Ans. Zein

Q.109 What is the protein molecule present in barley seed?

Ans. Hordein

Q.110 What is the name of salt soluble storage proteins present in legume seeds?

Ans. Globulins

Q.111 Mention the name of the storage protein in sunflower seeds?

Ans. Inulin

Q.112 What is the chromosome number of zygote?

Ans. 2n

Q.113 What is the seed part developed from zygote?
Ans. Embryo

Q.114 Name the crops that produce apomictic seed?
Ans. *Poa* sp., citrus

Q.115 Specify the crop which produces polyembryony seed?
Ans. Citrus, curry leaf

Q.116 Multigerm seeds can be seen in _________ crop.
Ans. Beetroot

Q.117 Which crop seed has folded cotyledons?
Ans. Cotton, bhendi

Q.118 ____________crop seeds having curved cotyledon
Ans. Tomato, brinjal, chilli, onion

Q.119 What is the name of the middle portion of monocot embryo?
Ans. Mesocotyl

Q.120 What is the name of middle part of a dicot embryo ?
Ans. Hypocotyl

Q.121 Name the structure that encloses the primary leaves in dicot seeds?
Ans. Plumule

Q.122 Give the name of the structure that covers the root in dicot seed?
Ans. Root cap

Q.123 What is the name of the spongy outgrowth in tapioca/castor seed?
Ans. Caruncle

Q.124 ___________tissue that develops into caruncle during seed development
Ans. Outer integument

Q.125 What is the role of caruncle in castor seed?

Ans. Absorbs water for seed germination

Q.126 What is meant by operculum?

Ans. Stopper like thick walled structure found over embryo in seeds. It imposes mechanical resistance to germination.

Q.127 What is the tissue which develops into operculum?

Ans. Inner integument

Q.128 Mention the seed which is having operculum?

Ans. Date palm, Victoria lily

Q.129 Give the name of some seed appendages?

Ans. Awn, caruncle, operculum, aril, mace, wings, fuzz

Q.130 What is the soft juicy coat surrounding the seed?

Ans. Aril

Q.131 Provide an example of edible arillate seed?

Ans. Pomegranate

Q.132 What is the name of the red coloured flesh present on the nutmeg seed?

Ans. Mace

Q.133 Provide an example of aril with medicinal property?

Ans. Mace of Nutmeg

Q.134 Cotton seed has the appendage called____________

Ans. Fuzz

Q.135 The appendage, wing is present in which crop seed?

Ans. Moringa

Q.136 Name the tissue that develops the aril on the seed.

Ans. Funiculus

Q.137 What is the name of the appendage in cocklebur seed?

Ans. Hooks

Q.138 What is the role of operculum?

Ans. It protects the seed against water imbibition due to its impervious nature.

Q.139 Function of aril / mace in a seed is __________

Ans. Supply of nutrients to the seed.

Q.140 Pericarp is hard and leathery and remains free from seed coat in ________ fruit.

Ans. Achene

Q.141 ________ is a dry indehiscent one seeded fruit with hard and woody pericarp

Ans. Cashewnut

Q.142 Name a crop where hard seed is included in germination per cent?

Ans. Blackgram

Q.143 Oil like substance present in the mesocarp of cashewnut is ____________

Ans. Cashew Nut Shell Liquid (CNSL)

Q.144 Liquid endosperm is present in ______ seed.

Ans. Coconut

Q.145 What is meant by "Putamen" ?

Ans. Leathery cartilaginous covering of a seed.

Q.146 Which crop seed possess putamen ?

Ans. Mango

□□□

Chapter 2

Pollination and Seedset

Q.1 ________refers to the transfer of pollen grains from anthers to ovule.

Ans. Pollination

Q.2 Sexual reproduction involves fusion of ________and ______ gametes.

Ans. Male and female

Q.3 Parthenocarpic fruit means ________

Ans. Fruit development without fertilization

Q.4 Female gametophyte is seven celled and __________nucleate structure

Ans. Eight

Q.5 Fusion of two polar nuclei and one sperm nuclei results in development of _________ in seed

Ans. Endosperm

Q.6 Fusion of one male nuclei with egg results in the formation of________ in seed.

Ans. Zygote / Embryo

Q.7 What is the mechanism of Cleistogamy?

Ans. Pollination occurs without opening of flowers. Eg. Bengal gram

Q.8 Pollination occurs after flower opening is known as ____________

Ans. Chasmogamy; eg. Wheat, Paddy, Oats, Barley.

Q.9 Chasmogamy is common in ________ pollinated crops

Ans. Self

Q.10 Allogamy refers to________

Ans. Cross pollination

Q.11 Xenogramy means _________

Ans. Cross pollination

Q.12 Autogamy is known as _______

Ans. Self pollination

Q.13 Self-incompatibility favours ___________pollination in crops

Ans. Cross

Q.14 Self-incompatibility is common in ___________crops

Ans. Sunflower, mustard, cabbage, cauliflower, radish

Q.15 What is meant by Dicliny?

Ans. Separation of male and female flowers in different parts on the same plant.

Q.16 What is the name of plant in which staminate and pistillate flowers are present in the same plant?

Ans. Monoecious

Q.17 Male and female flowers occur in different plants are called as_______

Ans. Dioecious

Q.18 What is meant by Dichogamy?

Ans. In a single flower of hermaphrodite nature, the stamens and pistils mature at different times.

Q.19 Flower pistils mature before stamen is called as__________

Ans. Protogyny

Q.20 The mechanism in which the anthers mature before pistil is called as_______

Ans. Protandry

Q.21 Tripping mechanism is common in___________ crop

Ans. Lucerne (alfalfa)

Q.22 The flower opening by tripping mechanism is done by________ in alfalfa.

Ans. Insects, particularly by alfalfa bees

Q.23 Anemophily is the mechanism of________

Ans. Wind pollination

Q.24 Entomophily is called as ________pollination in crops

Ans. Insect

Q.25 Pollination by water is named as ___________.

Ans. Hydrophily

Q.26 Pollination by animals is called as _______.

Ans. Zoophily

Q.27 Pollination by beetles is known as __________

Ans. Canthrophily

Q.28 Pollination by bats is known as __________

Ans. Chiropterophily

Q.29 Pollination by ants is known as ____________

Ans. Myrmecophily

Q.30 Pollination by birds is known as _________

Ans. Ornithophily

Q.31 What is meant by an electric bee?

Ans. Pollen collecting machine / Device for the collection of pollen

Q.32 What is the ideal temperature requirement for the activities of pollinators during seed production?

Ans. 24-38°C

Q.33 The absence of functional pollen grains in a hermophrodite flowers is termed as _________

Ans. Male sterility

Q.34 What is meant by masting?

Ans. Combined flowering and fruit bearing of trees in a forest

Q.35 What is geitonogamy?

Ans. Pollination occurring between flowers of same plant

□□□

Chapter 3

Seed Dispersal

Q.1 **Movement or transport of seed from one place to another place or away from the parent plant is called as __________**

Ans. Dispersal

Q.2 Seed dispersal by environmental influence is termed as __________

Ans. Serotiny

Q.3 What is seed dispersal by gravity called?

Ans. Clitochory

Q.4 Provide an example for clitochory

Ans. Coconut

Q.5 What is the name of the self seed dispersal?

Ans. Autochory

Q.6 Examples for autochory are__________

Ans. Balsam, touch-me-not, geranium

Q.7 Dispersal of seeds by wind is called as _________

Ans. Anemochory

Q.8 Examples for seed dispersal by wind are______

Ans. *Calotropis, Tridax sp., Taraxacum* sp., maples.

Q.9 Seed dispersal by animals is termed as ________

Ans. Zoochory

Q.10 What is the name of the seed dispersal by animal guts?

Ans. Endozoochory

Q.11 Seed dispersal on the outside of animals are called as __________

Ans. Epizoochory

Q.12 What is ectozoochory?

Ans. Seed dispersal through attachment on animals external surface

Q.13 Examples for epizoochorous seed dispersal are________

Ans. *Trifolium angustifolium, Achyranthus aspara,* cocklebur, datura

Q.14 What is the name of the seed dispersal by ants?

Ans. Myrmecochory

Q.15 What is the part present in the myrmecochorous plant seeds?

Ans. Elaiosome, a lipid-rich attachment which attracts ants

Q.16 Seed dispersal by water is termed as ________

Ans. Hydrochory

Q.17 Examples for seed dispersal by water are _______

Ans. Coconut, water lily, mangrove trees

Q.18 Characteristics of endozoochorous seed are ______

Ans. Hard seed coat, palatable fruit

Q.19 Characteristics of epizoochorous seed

Ans. Hook like appendages to cling or attach to animals

Q.20 Characteristics of hydrochorous seeds

Ans. Light and fibrous seed filled with air

Q.21 Characteristics of anemochorous seeds

Ans. Seeds are lighter with hair / wing like appendages

Q.22 What is samara?

Ans. Samara is a winged seed for wind dispersal

Q.23 Provide examples for samara.

Ans. Pine, mapple, moringa

□□□

Chapter 4

Seed Dormancy and Germination

Q.1 What are all the types of seed germination?

Ans. 1. Epigeal germination

2. Hypogeal germination

Q.2 What is the name of the germination where the cotyledon is under the ground?

Ans. Hypogeal

Q.3 What is meant by epigeal?

Ans. Epicotyl elongates and cotyledon comes above the soil surface during germination.

Q.4 Give examples for hypogeal germination?

Ans. All monocots, onion, redgram, chick pea

Q.5 What are all the crops that posses epigeal germination behavior?

Ans. All dicots except peas, bengal gram, redgram, rubber

Q.6 In monocot seeds _______part of the seed comes first during germination

Ans. Coleorhiza

Q.7 What is the level of polyamines during seed ageing or stress or low vigour seeds?

Ans. High

Q.8 _________enzymes involved during seed germination

Ans. α-amylase, β-amylase, maltase, proteinase

Q.9 What is the name of the water absorption phenomenon in seed?

Ans. Imbibition

Q.10 Whether dead seeds absorb water or not?

Ans. Absorb

Q.11 Which seed absorb water quickly, dead or fresh?

Ans. Dead seeds

Q.12 __________ part of the seed elongates first while germination in dicot seed?

Ans. Radicle

Q.13 What is the reason for swelling of seeds during imbibition?

Ans. Water imbibing proteins, cellulose and pectin compounds.

Q.14 What is the name of the seed that requires light for germination?

Ans. Photoblastic seeds

Q.15 Give an example for photoblastic seed?

Ans. Tobacco, Lettuce

Q.16 What is the name given to seeds that require darkness for germination?

Ans. Skotoblastic seeds

Q.17 Name some seeds that require darkness for germination?

Ans. Tomato, onion, datura

Q.18 What is the name of the inactive period of seed embryo?

Ans. Dormancy

Q.19 The dormancy in rice seed is caused by ___________and ___________

Ans. Hull; pericarp

Q.20 What is the water requirement level for seed germination?

Ans. 40% to 70% of dry weight basis

Q.21 What is the effect of alterin compound on seed?

Ans. It inhibits seed germination

Q.22 The name of the dormancy due to hard seed coat is called__________

Ans. Physical dormancy

Q.23 What is the main cause of hard seed coat?

Ans. Scelerification of seed coat surface especially legumes that makes it impermeable to water or gas

Q.24 Give examples for hard seed coat dormant seeds?

Ans. Blackgram, greengram (legume – pulses), lucerne, desmanthus (legume – forage), *Samania saman, Delonix regia* (legume – forest tree)

Q.25 Give an example for the seed that requires oxygen for germination?

Ans. Cocklebur seeds

Q.26 What is the name of the physical dormancy breaking method?

Ans. Scarification

Q.27 What are the acids used for chemical dormancy breaking / scarification?

Ans. Sulphuric acid, hydrochloric acid

Q.28 Give the methods for physical scarification in dormant seeds?

Ans. Seed coat rubbing, pricking, punching, piercing and nicking

Q.29 What is the machine used for mechanical scarification?

Ans. Scarifier

Q.30 What is meant by Bioscarification?

Ans. Scarification of seed coat by any biological agents eg. Passage through animal gut or soaking in cowdung solution.

Q.31 __________is prevented due to dormancy.

Ans. Pre-harvest sprouting / Precocious germination

Q.32 The endocarp is responsible for preventing the entry of water in __________

Ans. Mango

Q.33 What is the main cause for physiological dormancy in seeds?

Ans. Presence of inhibitors in seeds

Q.34 What is the main inhibitor present in the seed?

Ans. Abscissic acid (ABA)

Q.35 Give examples for dormancy in seeds due to ABA?

Ans. Cotton, groundnut

Q.36 Give an example for the seed embryo that contains inhibitor?

Ans. Cocklebur seed

Q.37 What is the physiological inhibiting substance present in sunflower seed?

Ans. Hydrocyanic acid

Q.38 The inhibitor coumarin is present in which crop seed?

Ans. Coriander

Q.39 What is the name of the inhibitor present in *Brassica* sp.?

Ans. Phenyl isothiocyanate

Q.40 Dormancy that sets on the mother plant is called as____________

Ans. Primary dormancy

Q.41 What is the name of the induced dormancy in seed?

Ans. Secondary dormancy

Q.42 What is the name of the dormancy induced in seed due to high temperature?

Ans. Thermo dormancy

Q.43 Example for thermo dormant seed?

Ans. Lettuce seed.

Q.44 What is the name of the dormancy induced in seed due to light?

Ans. Photo dormancy

Q.45 Give example for photo dormant seeds?

Ans. Oats, onion, amaranthus, lettuce

Q.46 What is the counteracting chemical for ABA inhibitor in seed?

Ans. Gibberellic acid

Q.47 What are the chemicals used for breaking dormancy in light requiring seeds?

Ans. Potassium nitrate, thiourea

Q.48 What is the simple method to remove the inhibitors from seed?

Ans. Soaking in water or leaching in running water.

Q.49 Which chemical is used to break physiological dormancy in groundnut and sunflower?

Ans. Ethrel

Q.50 What is the concentration of ethrel for groundnut and sunflower seed dormancy breaking?

Ans. 200 ppm

Q.51 What is the concentration of KNO_3 for physiological dormancy removal?

Ans. 0.2%

Q.52 What is the status of the embryo in physiologically dormant seed?

Ans. Fully developed and dormant

Q.53 What is the name of the dormancy due to under developed embryo in seeds?

Ans. Morphological dormancy

Q.54 Name the treatment which is used for morphologically dormancy removal?

Ans. Stratification

Q.55 What is the name of the seed storage period for removing the morphological dormancy?

Ans. After ripening period

Q.56 The changes that occur in dormant seeds after dispersal are called as ______

Ans. After ripening

Q.57 What is the name of the phenomenon that necessitates seeds to be exposed to low temperature for germination?

Ans. Low temperature stratification/cold stratification

Q.58 Provide some examples of seeds that require high temperature/ warm stratification for breaking seed dormancy?

Ans. Oil palm, lucerne, groundnut, sorghum, paddy

Q.59 What is the temperature requirement for cold stratification?

Ans. 0-5°C

Q.60 Give examples for crops that require stratification treatment?

Ans. Apple, pear, peach, plum

Q.61 What is the temperature requirement for hot water treatment in chickpea seeds?

Ans. 80°C for 1-5 minutes

Q.62 Treatment given to break seed dormancy in rice is _______.

Ans. Warm stratification

Q.63 Name the growth regulator for physiological dormancy removal other than gibberellic acid?

Ans. Kinetin

Q.64 What is the name of the inactive stage of seed due to unfavourable conditions?

Ans. Quiescence

Q.65 What is the name of the seed germination which occurs without dormancy on mother plant?

Ans. Viviparous germination

Q.66 What is the phenomeaon of seed that germinates on mother plant upon wetting by rain known as _________

Ans. Precocious germination

Q.67 In situ germination is termed as __________

Ans. Vivipary

Q.68 Give examples for vivipary seed germination?

Ans. Chow-chow (*Sechium edule*), Mangrove species (*Rhizophora apiculata, Avicennia marina*)

Q.69 What is the main advantage of seed dormancy?

Ans. Prevents premature germination

Q.70 In which crop the water stress induces secondary dormancy in seeds?

Ans. Carrot

Q.71 The presence of different types of seeds stored in soil is known as ?

Ans. Soil seed bank

Q.72 The phenomenon of seeds regenerating from soil seed bank is known as ________

Ans. Reseeding

□□□

Chapter 5

Seed Production and Quality

Q.1 What are the goals of seed technology?

Ans. a. Rapid multiplication
b. Timely supply
c. High seed quality
d. Reasonable price

Q.2 Expand the term IMSCS?

Ans. Indian Minimum Seed Certification Standards.

Q.3 As per Seeds Act 1966, the seed certification is ______and labeling is______

Ans. Voluntary, compulsory.

Q.4 Seed certification is an __________operation of quality control

Ans. Legal

Q.5 State level organization empowered to conduct seed certification work is ______

Ans. Seed Certification Agency

Q.6 How many phases are there in seed certification ?

Ans. Six phases

Q.7 Certification is done for__________ varieties alone.

Ans. Notified

Q.8 Who notifies the variety for seed production?

Ans. Central Variety Release Committee

Q.9 What is the maximum area that can be registered under a sowing report ?

Ans. 25 Acres

Q.10 LFR is the term used for _________

Ans. Liable For Rejection

Q.11 What is the validity period of a certification tag?

Ans. 9 months

Q.12 What is the re-validation period for certified seed?

Ans. 6 months

Q.13 Expand "GOT"?

Ans. Grow Out Test

Q.14 ODV is called?

Ans. Other Distinguishable Variety

Q.15 Table examination of seeds is otherwise called as _________

Ans. Other Distinguishable Variety (ODV) test

Q.16 NDUS______?

Ans. Novelty, Distinctness, Uniformity and Stability

Q.17 Expand VCU_________?

Ans. Value for Cultivation and Use

Q.18 SRR is abbreviated from ________

Ans. Seed Replacement Rate

Q.19 SRR is calculated as ____________

Ans. SRR = (X / Y) x 100

where, X = quantity of quality seed of a variety used in a year

Y = quantity of seed required to cover total area

Q.20 What is the seed replacement rate for hybrids?

Ans. 100%

Q.21 SMR is expanded as __________

Ans. Seed Multiplication Ratio

Q.22 Seed multiplication ratio means__________

Ans. Number of seeds multiplied from a single seed.

Q.23 Which crop has the lowest seed multiplication ratio?

Ans. Groundnut

Q.24 Seed multiplication ratio for groundnut is______

Ans. 1:8

Q.25 What is the seed multiplication ratio for rice?

Ans. 1:80

Q.26 What is the seed multiplication ratio for wheat?

Ans. 1:20

Q.27 In cereals which crop has the highest SMR and how much?

Ans. Pearlmillet, 1:200

Q.28 What is the SMR for major pulses like greengram, blackgram and cowpea?

Ans. 1:40

Q.29 Seed multiplication ratio for sesame is__________

Ans. 1:250

Q.30 What is the SMR for cotton varieties and hybrids?

Ans. Varieties - 1:50
Hybrids - 1:250 (delinted); 1:300 (fuzzy)

Q.31 Among the major vegetables, which crop has the highest SMR and how much?

Ans. Brinjal, 1:420-450

Q.32 Among vegetables which crop has the lowest seed multiplication ratio ?

Ans. Peas (1 : 10)

Q.33 AVT is the abbreviation of __________

Ans. Advanced Varietal Trial

Q.34 What is meant by VRC?

Ans. Variety Release Committee

Q.35 CVRC is the abbreviation of __________

Ans. Central Variety Release Committee

Q.36 SVRC__________?

Ans. State Variety Release Committee

Q.37 Seed replacement rate for rice is __________

Ans. 29%

Q.38 SRR for sorghum?

Ans. 23%

Q.39 SRR for pearlmillet?

Ans. 26%

Q.40 SRR for maize?

Ans. 30%

Q.41 SRR for wheat?

Ans. 19%

Q.42 SRR for black gram, green gram, cowpea is ______

Ans. 20%

Q.43 SRR for hybrid varieties?

Ans. 100%

Q.44 SRR for cotton varieties?

Ans. 15%

Q.45 What is the level of yield improvement when quality seed used?

Ans. 20%

Q.46 The seeds obtained from the single plant is called ________

Ans. Primordial seed

Q.47 Genetic purity of primordial seed is ________?

Ans. 100%

Q.48 Progeny of primordial seed is called ________?

Ans. Nucleus seed

Q.49 What are all the different classes of seed?

Ans.
1. Breeder seed
2. Foundation seed
3. Certified seed
4. Truthfully labelled seed

Q.50 The different seed production stages are called as ________ system of seed multiplication

Ans. Generation

Q.51 What is meant by three generation model?

Ans. Breeder seed → Foundation seed → Certified seed

Q.52 Four generation model of seed production is followed as_____

Ans. Breeder seed → Foundation seed I → Foundation seed II → Certified seed

Q.53 The number of generations a seed can be multiplied beyond breeder seed is ____

Ans. 3

Q.54 What is wrong with this seed production programme: BS-FSI-FSII-CSI-CSII

Ans. There cannot be more than three generations beyond BS

Q.55 What follows Foundation Seed Stage II?

Ans. Certified Seed

Q.56 In which crop, four or five generation model of seed production is allowed?

Ans. Highly self pollinated crops

Crops having low seed multiplication ratio

Q.57 Examples for four generation seed multiplication model crops are _________

Ans. Groundnut, rice

Q.58 NS is the abbreviation of __________

Ans. Nucleus seed

Q.59 What is meant by BS?

Ans. Breeder seed

Q.60 FS means___________

Ans. Foundation seed

Q.61 RS is expanded as________

Ans. Registered seed

Q.62 What is the other name of foundation seed in European countries?

Ans. Registered seed

Q.63 What is meant by CS?

Ans. Certified seed

Q.64 TFL is expanded as________

Ans. Truthfully Labelled Seed

Q.65 Breeder seed is the progeny of _______seed.

Ans. Nucleus seed

Q.66 Progeny of the breeder class seed is________

Ans. Foundation seed

Q.67 Certified seed is produced from ________ class seed

Ans. Foundation

Q.68 Does nucleus seed have tag or not?

Ans. No

Q.69 The colour of the tag for breeder seed is________

Ans. Golden Yellow (ISI 5-1978)

Q.70 The colour of tag for foundation class seed is________

Ans. White

Q.71 The certified seed should be attached with ________ colour tag.

Ans. Azure blue (ISI No.104)

Q.72 What it the colour of certification tag stage II?

Ans. Azure Blue

Q.73 The colour of tag for truthfully labelled seed is ______

Ans. Opal Green

Q.74 The production of nucleus seed and breeder seed is otherwise called as ___

Ans. Maintenance Breeding

Q.75 Maintenance methods of seed quality were prescribed by

Ans. Hartmann and Kester, 1968

Q.76 __________production is exempted from certification by certification agency

Ans. Breeder seed

Q.77 Minimum genetic purity requirement for nucleus seed and breeder seed is __?

Ans. 100%

Q.78 Genetic purity of foundation class seed is ____?

Ans. 99%

Q.79 Genetic purity of certified class seed is _________?

Ans. 98%

Q.80 What is the minimum genetic purity requirement for hybrids?

Ans. 95%

Q.81 Which crop hybrid requires 90% genetic purity?

Ans. Cotton, tomato, brinjal, true potato seed, watermelon

Q.82 What is the minimum genetic purity requirement for castor hybrids?

Ans. 85%

Q.83 What is the minimum genetic purity level of cotton, tomato, brinjal, melon hybrids?

Ans. 90%

Q.84 The lowest genetic purity percentage of 85% is allowed for foundation and certified seeds of ______crop

Ans. Castor

Q.85 Which class seed comes under the purview of seed certification agency?

Ans. Foundation and certified seeds

Q.86 What is the favourable temperature requirement during vegetative stage in cole crops seed production?

Ans. 4-10°C

Q.87 Cole crops require temperature of __________at reproductive stage for seed development and maturation.

Ans. 15-20°C

Q.88 Varietal deterioration occurs due to the factors like ________

Ans. Natural crossing, mechanical admixtures, developmental variation, mutation, minor genetic variation

Q.89 The most important source of varietal deterioration is ____________

Ans. Natural crossing

Q.90 Pollen shedder causes ____________contamination in hybrid seed production field.

Ans. Genetic

Q.91 Can varieties be grown outside their area of adaptation for higher yields?

Ans. No. It leads to faster deterioration of varieties.

Q.92 Developmental variations in seed crops is due to ____________

Ans. Location

Q.93 Volunteer plants in seed production field comes from____________

Ans. Self sown seed of previous crop

Q.94 Removal of unwanted plant in the seed field is called as ___________

Ans. Roguing

Q.95 Rogue spacing for paddy foundation and certified seeds are ________and __________respectively.

Ans. 30 cm for every 4 feet; 30 cm for every 8 feet

Q.96 Process of evaluating the seed crop in the field is termed as ___________

Ans. Field inspection

Q.97 'Rabbit ear' in peas is due to__________

Ans. Mutation

Q.98 High temperature in pearlmillet causes__________ symptom during flowering.

Ans. Tip sterility

Q.99 Boron deficiency causes 'hollow heart' symptom in ________ crop seed.

Ans. Groundnut

Q.100 What is the symptom of calcium deficiency in groundnut seed?

Ans. Darkened plumule

Q.101 The hard seed development in pulses is mainly due to ________ temperature during seed development and maturity.

Ans. High

Q.102 'Marsh spot' in peas is caused by _____deficiency

Ans. Manganese

Q.103 'Whiptail' symptom is common in ________ crop which is due to __________ deficiency.

Ans. Cauliflower, molybdenum

Q.104 What is the name of the germination due to high rainfall at harvest?

Ans. Insitu germination or precocious germination

Q.105 Insitu germination due to high rainfall is common in ________ crops

Ans. Groundnut, sorghum, pearlmillet, rice

Q.106 Low relative humidity causes ________ development in seeds.

Ans. Hard seed

Q.107 High temperature causes ________ in tomato which leads to flower drop.

Ans. Stigma elongation out of the anther cone

Q.108 High rainfall causes ________ seeds in ________ crops.

Ans. Off coloured; pulse

Q.109 'Mottled seeds' can be seen in ________ crop seeds.

Ans. Pea

Q.110 In which crop aerated steam therapy is followed?

Ans. Sugarcane

Q.111 In which crop do you encounter bleached seed?

Ans. Peas

Q.112 Flower opening in a stalk towards upward is termed as ______

Ans. Acropetal

Q.113 Flowering towards centre of the inflorescene is known as ________

Ans. Centripetal

Q.114 Flowering in a stalk towards downward is called as ________

Ans. Basipetal

Q.115 Give examples for basipetal flowering

Ans. Rice, sorghum

Q.116 What is the name for the separation of seed crop by other contaminant crops or weeds?

Ans. Isolation

Q.117 Separation of seed crop and contaminant sources by giving some prescribed distance is called as ________

Ans. Isolation distance

Q.118 Time isolation is practiced in ________crops

Ans. Self pollinated

Q.119 Sowing in different dates to prevent contamination is otherwise called as ________

Ans. Time isolation

Q.120 ______ isolation is nothing but raising the seed crop in different altitudes to avoid contamination.

Ans. Physiological

Q.121 Physiological isolation is given for ________ crops

Ans. Temperate

Q.122 Seeds should be sown in the depth of ________

Ans. 6 cm

Q.123 Generally the seeds should be sown in the soil based on ______the size of the seed.

Ans. Double

Q.124 Growing of plants all around the field to avoid foreign pollen contamination is called ________

Ans. Live Barrier Isolation

Q.125 Staggering is followed in ________ seed production

Ans. Hybrid

Q.126 What is the name of the hybrid when two parents of a species are crossed?

Ans. Intervarietal hybrid or intraspecific hybrid

Q.127 What is meant by distant hybridization?

Ans. It is the cross between different species of the same genus or of different genera

Eg. Interspecific hybrid, Intergeneric hybrid

Q.128 Crossing of entirely different species of the same genus is called as________

Ans. Interspecific hybrid

Q.129 Crossing of two species belonging to two different genera is termed as______

Ans. Intergeneric hybrid

Q.130 Crossing of two inbreds is known as ______

Ans. Single cross hybrid

Q.131 What is meant by three way cross?

Ans. Crossing of one single cross hybrid with an inbred

Q.132 What is the term called when one single cross hybrid is crossed with an open pollinated variety?

Ans. Double top cross hybrid

Q.133 How many inbreds are required in double cross hybrid?

Ans. Four inbreds

Q.134 Maintainer lines are not necessary in ________ line breeding system

Ans. Two

Q.135 What is meant by GMS system?

Ans. Geneic Male Sterility

Q.136 CGMS is called as ________

Ans. Cytoplasmic Geneic Male Sterility system

Q.137 What are unobserved rows during field inspection known as?

Ans. Cross over rows

Q.138 What is meant by EGMS?

Ans. Environmental Sensitive Genic Male Sterility

Q.139 TGMS is expanded as ________

Ans. Temperature Sensitive Genic Male Sterility

Q.140 What is meant by PGMS?

Ans. Photoperiod Sensitive Genic Male Sterility

Q.141 In which crop the PGMS/TGMS systems are used for hybrid seed production?

Ans. Rice

Q.142 What is meant by Gametocides / Chemosterilants?

Ans. The chemicals used to create male sterility

Q.143 What is meant by CHA's?

Ans. Chemical Hybridizing Agents

Q.144 Examples for CHA's are ________

Ans. Malic hydrazide (MH), Ethrel, Sodium methyl arsenate, Zinc methyl arsenate.

Q.145 What is meant by three line breeding system in hybrid seed production?

Ans. Use of three lines as parental materials viz., A, B and R lines in CGMS system

Q.146 Example for three line breeding system in hybrid seed production is _________.

Ans. Cytoplasmic genic male sterility

Q.147 Methods involved in two line breeding system in hybrid seed production are ___________.

Ans. Emasculation and dusting, detasseling and pollination, genetic male sterility.

Q.148 Methods involved in single line breeding system are _______.

Ans. Apomixes, tissue culture

Q.149 What is the name of the 'A' line in cytoplasmic male sterility system?

Ans. Female parent or male sterile parent

Q.150 Which line is isogenic of 'A' line in cytoplasmic male sterility system of hybrid seed production?

Ans. B line or male fertile parent

Q.151 What is the other name of 'B' line in cytoplasmic male sterility system?

Ans. Maintainer line or non-pollen restorer male line

Q.152 What is the name of the 'R' line in cytoplasmic male sterility system of hybrid seed production?

Ans. Restorer line or male parent

Q.153 What are all the stages/generations used for the multiplication of parental lines in hybrid seed production using CGMS system?

Ans. Breeder and foundation seed production stages

Q.154 Flowering of male and female parents uniformly is called as _____________

Ans. Synchronization

Q.155 Synchronization is also known as__________

Ans. Nicking

Q.156 Adjustment of sowing dates of parental line seeds for getting synchronized flowering in hybrid seed production is termed as________

Ans. Staggering or staggered sowing

Q.157 Presence of B line plants in A line in CGMS system is called ___________

Ans. Pollen shedders

Q.158 Presence of male parent (R line) in female rows (A line) during CGMS based hybrid seed production is called ______

Ans. Rogue

Q.159 What is the presence of 'A' line in 'B' line called ?

Ans. Off type

Q.160 What is the presence of 'R' line in 'B' line known as ?

Ans. Rogue

Q.161 Male plants raised all sides of the field in hybrid seed production plot are called as __________

Ans. Border rows

Q.162 How many border rows are required for varietal seed production?

Ans. None

Q.163 The differentiation of male and female parents can be done by providing the ____in the hybrid seed production fields

Ans. Markers

Q.164 Give examples for live marker?

Ans. Redgram, sunflower

Q.165 Give examples for dead marker?

Ans. Painted stake, ribbon, coloured plastic tag

Q.166 What is the main pollinator in seed production?

Ans. Bees

Q.167 Name the pollinator used in alfalfa seed production?

Ans. Alkali bee (*Nomia melanderi*)

Q.168 What is the term for bee pollination?

Ans. Melitophily

Q.169 How will you improve seed set in dioecious crop?

Ans. Bee pollination or apiary

Q.170 Provide examples for anemophilous crop?

Ans. Bajra, rice

Q.171 Can a crop in which 1/3 area has been lodged accepted for certification?

Ans. No

Q.172 Can a rice crop without rouging space accepted for certification?

Ans. No

Q.173 Does mixed cropping or intercropping is allowed for seed production ?

Ans. No

Q.174 In hybrid seed production __________ parent is harvested first.

Ans. Male

Q.175 Wheat is a ________ pollinated crop

Ans. Self

Q.176 Isolation requirement for wheat is ________ m

Ans. 3

Q.177 Hybrid seed production method used in wheat is __________

Ans. Cytoplasmic genic male sterility

Q.178 Egg floatation technique is used for _________ seed upgradation

Ans. Rice

Q.179 Who is the father of hybrid rice ?

Ans. Yuan Longping

Q.180 What is the objectionable weed species in rice?

Ans. Wild rice (*Oryza sativa* var. *fatua*)

Q.181 What is the period of time isolation in rice seed crop?

Ans. 21 days

Q.182 What is the isolation distance required for foundation and certified classes of seed production in rice varieties?

Ans. 3 m

Q.183 Rice foundation stage requires _______m isolation distance for hybrid seed production.

Ans. 200

Q.184 Isolation distance required for paddy certified hybrid seed production is_____

Ans. 100 m

Q.185 Planting ratio for rice hybrid seed production is _________

Ans. Female : Male - 8:2

Q.186 Number of border rows for rice hybrid seed production is _________.

Ans. 4 rows

Q.187 What is the favourable temperature and relative humidity for high seedset in rice hybrid seed production?

Ans. Temperature: 24-30°C Humidity : 70- 80%

Q.188 What is meant by staggered sowing?

Ans. Adjustment of sowing date of parental lines to get synchronized flowering and higher seedset.

Q.189 Rope pulling method is followed in ________hybrid seed production

Ans. Rice

Q.190 In which crop do you follow rod driving for hybrid seed production ?

Ans. Rice

Q.191 What is the name of the growth regulator used for panicle exertion in hybrid rice seed production?

Ans. Gibberellic acid

Q.192 Male sterile 'A' line used in hybrid seed production exhibits symptoms of fertility on a particular portion, either on only the main tiller or one or more of the side tillers is called as ________

Ans. Partials

Q.193 Maturity index of rice seed is____________

Ans. Change of seed colour to straw yellow

Q.194 Sorghum is a _______pollinated crop.

Ans. Often cross

Q.195 In sorghum the flowering starts from tip of the panicle and proceeds downward which is otherwise called as _______

Ans. Basipetal

Q.196 Sorghum requires _________m and _______m isolation distance for foundation and certified classes respectively.

Ans. 200, 100

Q.197 What is the chemical recommended for seed hardening in sorghum?

Ans. Potassium di hydrogen orthophosphate ($KH_2 PO_4$)

Q.198 Isolation distance for certified seed production of sorghum in the presence of Johnson grass in the vicinity

Ans. 400 m

Q.199 Johnson grass is a objectionable weed in ______crop.

Ans. Sorghum

Q.200 What is the isolation distance for sorghum hybrid seed production?

Ans. Foundation seed – 300 m
Certified seed – 200 m

Q.201 What is the tool used for sorghum hybrid seed production?

Ans. Cytoplasmic genic male sterility (CGMS)

Q.202 Planting ratio followed in sorghum hybrid seed production is ________

Ans. Female : Male – 4:2

Q.203 How many number of border rows are required for sorghum hybrid seed production?

Ans. 4 rows

Q.204 What is the pollination behaviour of pearlmillet crop?

Ans. Cross pollination

Q.205 The mechanism for cross pollination in pearlmillet is _______.

Ans. Protogyny

Q.206 What is meant by jerking?

Ans. Removal of advanced flowering ear-head

Q.207 In which crop jerking is followed?

Ans. Pearlmillet

Q.208 What is the isolation distance requirement for pearlmillet varietal seed production?

Ans. Foundation seed – 400 m; Certified seed – 200 m

Q.209 What is the simple technique used to remove the ergot affected seeds in a seedlot of sorghum or pearlmillet?

Ans. Salt floatation technique

Q.210 __________is the method of hybrid seed production in pearlmillet?

Ans. Cytoplasmic genic male sterility

Q.211 The isolation distance requirement for the foundation and certified classes of hybrid seed production in pearlmillet are __________

Ans. Foundation seed – 1000 m; Certified seed – 200 m

Q.212 What is the planting ratio for pearlmillet hybrid seed production?

Ans. 6 (female) : 2 (male)

Q.213 How many number of border rows are recommended for pearlmillet hybrid seed production?

Ans. 8

Q.214 Among the crops like rice, sorghum and pearlmillet, which crop has the less problem of synchronization during hybrid seed production?

Ans. Pearlmillet

Q.215 Which factor favours reduced problem of synchronization during hybrid seed production in pearlmillet?

Ans. Tillering habit

Q.216 Why jerking is advised in pearlmillet crop during seed production?

Ans. To get synchronized and uniform flowering

Q.217 What is the layer formed in sorghum, pearlmillet and maize crops during the attainment of the seed physiological maturity?

Ans. Dunken layer

Q.218 Mode of pollination in maize is called__________

Ans. Geitonogamy

Q.219 What is the name of the female reproductive part in maize?

Ans. Cob

Q.220 Indicate the name of the male reproductive part in maize.

Ans. Tassel

Q.221 What is the name of the maize style?

Ans. Silk

Q.222 Which part of the style is receptive in a maize plant?

Ans. Throughout its length (about 32 cm)

Q.223 How many pollens are produced by a normal maize plant?

Ans. Approximately 25,00,000 pollens

Q.224 The mechanism favouring the cross pollination in maize is__________

Ans. Protandry

Q.225 The method followed in maize hybrid seed production is __________

Ans. Detasselling and pollination

Q.226 What is the number of border rows required for maize single cross hybrid seed production?

Ans. 4 rows

Q.227 Composite / synthetic varieties are developed in ________crop.

Ans. Maize

Q.228 What are all the types of hybrids available in maize?

Ans. Single cross hybrid, double cross hybrid, three way cross hybrid

Q.229 What is the flowering behaviour of maize crop?

Ans. Monoecious

Q.230 What is the isolation distance requirement for maize synthetic varieties?

Ans. Foundation class – 400 m (same for inbreds, varieties and composites)

Certified class - 200 m

Q.231 The maize crop requires isolation distance of _______m for foundation class and __________m for certified class in hybrids with same kernel colour and texture as that of the seed parent

Ans. 400 m ; 200 m

Q.232 What is the isolation distance requirement for maize hybrids those differing in texture and kernel colour?

Ans. Foundation class – 600 m
Certified class – 300 m

Q.233 Planting ratio for maize single cross hybrid is ________

Ans. 4 (female) : 2 (male)

Q.234 Maize double cross hybrid requires ________rows of female and ______rows of male parents.

Ans. 6, 2

Q.235 What is the presence of tassel in female row of maize hybrid seed production called ?

Ans. Shedding tassel

Q.236 Cob sorting is followed in ________crop

Ans. Maize

Q.237 Which crop shows xenia and metaxenia effect in seed?

Ans. Maize

Q.238 Change in seed coat colour due to foreign pollen contamination leads to ____ effect.

Ans. Xenia

Q.239 The genetic contamination due to foreign pollen causes colour change in endosperm is termed as ________

Ans. Metaxenia

Q.240 What is the minimum germination requirement for most of the cereals and millets?

Ans. 80%

Q.241 What is the minimum physical purity requirement for most of the cereals and millets?

Ans. 98%

Q.242 What is the minimum germination requirement for maize hybrid seeds?

Ans. 90%

Q.243 What is the pollination behaviour of redgram crop?

Ans. Often cross pollination

Q.244 What is the main pollinating agent in redgram crop?

Ans. Honey bees

Q.245 Isolation distance requirement for redgram foundation and certified seed classes are______________

Ans. Foundation seed - 200 m

Certified seed - 100 m

Q.246 Tool used for hybrid seed production in redgram is __________

Ans. Genetic male sterility system

Q.247 Planting ratio for redgram hybrid seed production is ________

Ans. 6:1 or 4:1 (female:male)

Q.248 __________number of border rows are required for redgram hybrid seed production?

Ans. 2

Q.249 ________of redgram pods is done to induce more flowers in male parent for continuous supply of pollens to the female parent.

Ans. Clipping

Q.250 Rain at the time of harvest causes ________and ________seeds in redgram.

Ans. Off-coloured; dimpled

Q.251 The widely used material as foliar spray to increase seedset in pulses is ______

Ans. Di-ammonium phosphate (DAP)

Q.252 Pollination behaviour of blackgram and greengram crops is ___________

Ans. Self pollination

Q.253 Isolation distance requirement for blackgram and greengram crops is ______

Ans. Foundation seed – 10 m

Certified seed – 5 m

Q.254 What is the method used to prevent imbibitional injury in pulses during pre-sowing seed treatment?

Ans. Slow imbibition or slow hydration or moist conditioning

Q.255 Minimum germination percentage requirement for pulses is ________

Ans. 75%

Q.256 Minimum physical purity requirement for pulses is ________

Ans. 98%

Q.257 Tendril clipping is followed for ________seed production

Ans. Cowpea

Q.258 What is the pollination behaviour of groundnut?

Ans. Self pollination

Q.259 The mechanism favouring the self pollination in groundnut is ________

Ans. Stigma remain closed in the keel petal

Q.260 Isolation distance required for groundnut foundation seed production ______

Ans. 3 m

Q.261 How many generation systems are allowed for groundnut seed production? Why?

Ans. 5; highly self pollination

Q.262 Name the three types of varieties available in groundnut?

Ans. Bunch, semi-spreading and spreading types

Q.263 What is the chemical used during pre-sowing hardening treatment of groundnut seed to eliminate the calcium deficiency?

Ans. Calcium chloride (0.5%)

Q.264 The pegs penetration, pod formation and filling can be increased by the application of ____________in groundnut.

Ans. Gypsum (@ 400 kg ha^{-1})

Q.265 What are the physiological maturity indices in groundnut?

Ans. Drying and falling of older leaves

Yellowing of the tip leaves

Turning of inner side of pod colour to black

Rattling sound of the seeds in the pods.

Q.266 What is the name of the mechanism of removal of pod from the mother plant in groundnut?

Ans. Stripping

Q.267 What is the name of the machine used for stripping of groundnut pods?

Ans. Groundnut stripper

Q.268 What is the time of last inspection in groundnut?

Ans. After harvest i.e. pod verification or pod sorting

Q.269 Which machine is used to remove the seeds from the groundnut pods?

Ans. Groundnut decorticator

Q.270 What is the minimum physical purity requirement for groundnut?

Ans. 96%

Q.271 Minimum germination requirement for groundnut is ____________

Ans. 70%

Q.272 What is the safe moisture content for groundnut seed storage in vapour proof and pervious containers?

Ans. Vapour proof container – 5%
Vapour pervious container – 9%

Q.273 Sunflower is a ___________ pollinated crop.

Ans. Cross

Q.274 The floral inflorescence in sunflower is also called as ___________?

Ans. Head or thalamus or capitulam

Q.275 The outer most floret in sunflower disc is called as__________?

Ans. Ray flower (male flower)

Q.276 The inner part flowers in sunflower head are__________

Ans. Disc flowers (hermaphrodite flower)

Q.277 What mechanism stimulates cross pollination in sunflower?

Ans. Self incompatibility and Protandry

Q.278 What is the flowering pattern in sunflower?

Ans. Centripetal

Q.279 What is the main pollinating agent in sunflower?

Ans. Honey bees

Q.280 Varietal renovation method is used in __________crop

Ans. Sunflower

Q.281 What is the previous crop requirement for sunflower ?

Ans. One year

Q.282 What is the tool used for hybrid seed production in sunflower?

Ans. Cytoplasmic genic male sterility (CGMS)

Q.283 What is the planting ratio followed for sunflower hybrid seed production?

Ans. 4 : 1 (female : male)

Q.284 Isolation distance for sunflower varietal seed production in foundation and certified classes are________

Ans. Foundation class – 400 m
Certified class - 200 m

Q.285 The parental line multiplication stages should be isolated with the distance of ________m in sunflower.

Ans. 600

Q.286 Sunflower hybrid seed production fields should be separated by the distance of _________m.

Ans. 400

Q.287 The R line / male parent should be selected with ____________ head types for sunflower hybrid seed production.

Ans. Multiple

Q.288 In sunflower the removal of side heads during male parent multiplication stage is called __________

Ans. Nipping

Q.289 Nipping of terminal flower is followed for getting higher ____________ in sunflower male parent.

Ans. Floral numbers

Q.290 How the pollination is supplemented in sunflower?

Ans. Keeping bee hives

Spraying Neera solution (contains 20% sucrose, 5% minerals etc.) to attract insects thereby increases the pollination

Rubbing the ear head with muslin cloth

Sib mating

Q.291 What is the visual symptom of physiological maturity in sunflower ?

Ans. Change of thalamus colour from green to yellow

Q.292 Sesame is a __________pollinated crop.

Ans. Often cross

Q.293 Isolation requirement for sesame is__________

Ans. Foundation seed – 100 m

Certified seed - 50 m

Q.294 Classify the sesame varieties based on seed colour?

Ans. White seeded varieties

Brown seeded varieties

Black seeded varieties

Q.295 The seed rate for gingelly is______

Ans. 3-4 kg / ha

Q.296 The recommended plant density for better seed yield in sesame is _________

Ans. 11 plants / sq. m.

Q.297 The disease that changes the gingelly plant parts from reproductive to vegetative stage is__________

Ans. Phyllody

Q.298 What is the minimum physical purity percentage for gingelly?

Ans. 97%

Q.299 The germination requirement for gingelly is_________

Ans. 80%

Q.300 What is the name of the castor inflorescence?

Ans. Raceme

Q.301 The castor inflorescence has _______flowers at bottom and _______flowers at top.

Ans. Male; female

Q.302 The percentage of female flowers in a castor inflorescence is _________

Ans. 30-50%

Q.303 The mechanism that favours the cross pollination in castor is _________

Ans. Protogyny

Q.304 What are all the pollinating agents in castor?

Ans. Birds, insects

Q.305 What are all the basic sex forms in castor?

Ans.
1. Monoecious (M)
2. Pistillate (P)
3. Interspersed Staminate Flower (ISF)
4. Sex Revertant

Q.306 What is the temperature that alters the sex forms in castor?

Ans. 32°C

Below 32°C – induces femaleness

Above 32°C – induces maleness

Q.307 What is the name of the white waxy coating on castor plant?

Ans. Bloom

Q.308 Mention the different types of bloom in castor

Ans. 1. No bloom
2. Single bloom
3. Double bloom
4. Triple bloom

Q.309 Main advantage of the bloom on castor plant is ________

Ans. Natural protection against extreme weather conditions like cold and drought

Q.310 Tool used for castor hybrid seed production is __________.

Ans. Pistillate mechanism

Q.311 Which country released the first castor hybrid ?

Ans. India

Q.312 What is the name of the first castor hybrid?

Ans. GCH-1 (Gujarat Castor Hybrid 1)

Q.313 What is the isolation distance requirement for castor seed production?

Ans. Foundation seed - 300 m
Certified seed - 150 m

Q.314 What is the ideal season for castor varietal seed production?

Ans. Kharif or summer

Q.315 What is the ideal season for castor hybrid seed production?

Ans. Rabi or winter

Q.316 What is the planting ratio for castor hybrid seed production?

Ans. 6:2 or 3:1 (female : male)

Q.317 The number of field inspections required for castor seed production is _____

Ans. 4

Q.318 Moisture stress leads to production of more _____ flowers in castor

Ans. Male

Q.319 Pollination behavior of cotton is ___________

Ans. Often cross pollination

Q.320 Two types of cotton are_________

Ans. Asiatic type (diploid cotton)

American type (tetraploid cotton)

Q.321 Asiatic type includes the species viz.,________, ________

Ans. *Gossypium herbaceum*

Gossypium arboreum

Q.322 American cotton includes the species such as ________, ___________

Ans. *Gossypium hirsutum*

Gossypium barbadense

Q.323 Examples for G. *hirsutum* cotton varieties are _________

Ans. LRA 5166, Anjali, MCU 5, MCU 5 VT, Supriya

Q.324 Example for G. *barbadense* cotton variety is ______

Ans. Suvin

Q.325 Classify the cotton varieties based on the fiber length

Ans. Short stable

Medium stable

Long stable

Extra long stable

Q.326 Give an example of national variety in cotton?

Ans. LRA 5166

Q.327 Suvin is a __________stable variety in cotton.

Ans. Extra Long

Q.328 What is the first cotton hybrid in India?

Ans. H1 (CO2 x SeaIsland V 135)

Q.329 Who is the father of hybrid cotton in India?

Ans. Dr. C.T. Patel

Q.330 Parentage of Varalakshmi cotton hybrid are_______

Ans. Lakshmi x SB 289 E

Q.331 Jayalakshmi (DCH 32) hybrid in cotton is crosses between the parents_____

Ans. DS 28 x SB 425

Q.332 Give examples for intraspecific cotton hybrids?

Ans. Savitha, Surya

Q.333 Examples for interspecific cotton hybrids are__________

Ans. Varalakshmi, Jayalakshmi, Sruthi, TCHB 213

Q.334 What is the name of the GMS hybrid in cotton?

Ans. Suguna (Gregg x K 3400)

Q.335 What is the seed setting percentage in cotton hybrid seed production?

Ans. Diploid cotton - 25%
Tetraploid cotton - 40-45%

Q.336 What is the isolation distance for cotton seed production?

Ans. Foundation seed - 50 m
Certified seed - 30 m

Q.337 What system is used for commercial hybrid seed production in cotton?

Ans. Emasculation and dusting

Q.338 Who developed the emasculation technique in cotton for hybrid seed production?

Ans. Doak (1934)

Q.339 Emasculation means removal of ________ flowers in monoecious crops.

Ans. Male

Q.340 Emasculation should be done during previous day evening between ________ hours for hybrid cotton seed production

Ans. 2-6 PM

Q.341 Dusting of pollens in female flowers during cotton hybrid seed production should be done at ___________

Ans. 8 AM -12 Noon

Q.342 One male flower can be used to dust________female flowers in hybrid cotton seed production

Ans. 5

Q.343 Planting ratio for cotton hybrid seed production is ________

Ans. 4:1 (female : male)

Q.344 In which crop the topping operation followed for seed production?

Ans. Cotton

Q.345 What is meant by topping?

Ans. Nipping or removal of apical shoot

Q.346 What is the name of the growth regulator spayed in cotton for increasing the flower setting?

Ans. Naphthalene Acetic Acid (NAA) @ 40 ppm

Q.347 The tinge present at the base of cotton flower petal is called__________

Ans. Petal spot

Q.348 Petal spot is present in the cotton cultivated species viz.,______________

Ans. *Gossypium barbadense*
Gossypium herbaceum
Gossypium arboreum

Q.349 Reddening in cotton is due to ___________deficiency

Ans. Magnesium

Q.350 What is meant be 'hard locks' in cotton?

Ans. Kapas without proper bursting is called as hard locks

Q.351 Which crop the kapas sorting is followed?

Ans. Cotton

Q.352 Kapas sorting is the ____________in cotton.

Ans. Removal of off type and insect damaged bolls

Q.353 Ginning is the removal of _____from the seed.

Ans. Lint

Q.354 Delinting is the removal of __________on the seed.

Ans. Fuzz

Q.355 Two methods of delinting in cotton are______________

Ans. Acid delinting
Gas delinting

Q.356 ___________acid at the rate of ___________is used for cotton delinting.

Ans. Sulphuric
100 ml / kg of seed

Q.357 __________ machine is used for cotton seed upgrading.

Ans. Specific gravity separator

Q.358 What is the germination percentage requirement for cotton?

Ans. 65%

Q.359 Pollination behavior of tomato plant is ________

Ans. Self pollination

Q.360 Flower drop in tomato is due to ________

Ans. High temperature

Q.361 Self pollination in tomato is favoured by _______

Ans. Position of receptive stigma within the cone anthers
Pendant position of the flower

Q.362 Isolation distance requirement for tomato varietal seed production field is____

Ans. Foundation seed - 50 m
Certified seed - 25 m

Q.363 Tomato hybrid seed production field requires the isolation distance of ______and______ for foundation and certified classes.

Ans. 200 m ; 100 m

Q.364 Seedling age for transplanting in tomato is ___________

Ans. 25-30 days

Q.365 Hybrid seed production method followed in tomato is__________

Ans. Emasculation and dusting

Q.366 During field inspection _______, ________ and ________ disease affected plants should be removed in tomato

Ans. Early blight ; Leaf spot ; Tobacco mosaic virus (TMV)

Q.367 ________extraction method is used in tomato for extracting the seed.

Ans. Wet

Q.368 The methods used for extracting the seed in tomato are________

Ans. Fermentation method
Acid method
Alkali method

Q.369 What is the acid used for tomato seed extraction?

Ans. Hydrochloric acid @ 10 ml / kg of pulp

Q.370 ________alkali is used for tomato seed extraction.

Ans. Sodium bicarbonate (Na_2CO_3) @ 80 g / kg of pulp

Q.371 Seed recovery percentage in tomato is ....

Ans. 0.8-1.0%

Q.372 Brinjal is a ________ pollinated crop.

Ans. Often cross

Q.373 Heterostyly flowers are found in _______crop

Ans. Brinjal

Q.374 Mention the four different types of style in brinjal.

Ans.
1. Long style
2. Medium style
3. Short style
4. Pseudo-short style

Q.375 Production of __________ and __________style flowers is desirable for seed production and yield.

Ans. Long; medium

Q.376 What is the system used for hybrid seed production in brinjal ?

Ans. Emasculation and Dusting

Q.377 Isolation distance required for brinjal foundation and certified classes seed production is________

Ans. Foundation seed - 200 m
Certified class - 100 m

Q.378 Spraying of ___________reduces flower drop in brinjal

Ans. Naphthalene Acetic Acid (NAA) @ 20 ppm

Q.379 What are all the disease affected plants not allowed during field inspection in brinjal ?

Ans. Blight
Little leaf

Q.380 What is the designated disease in brinjal ?

Ans. Phomopsis blight

Q.381 The visual physiological symptom of maturity in brinjal is ___________

Ans. Change of fruit colour into yellow

Q.382 The seed recovery in brinjal is ___________

Ans. 5-6%

Q.383 What is the pollination behavior of chilli crop?

Ans. Cross pollination

Q.384 The system favours cross pollination in chilli is ________

Ans. Protogyny

Q.385 Isolation distance required for chilli foundation and certified seed production is ________

Ans. Foundation seed - 400 m
Certified seed - 200 m

Q.386 Age of the seedling for transplanting in chilli crop is ___________

Ans. 30 days

Q.387 Tool used for hybrid seed production in chilli is ________

Ans. Emasculation and Dusting

Q.388 Seed extraction methods used in chilli are________

Ans. Dry extraction
Wet extraction

Q.389 What is the seed recovery in chilli?

Ans. 40-50% of the dried fruit weight

Q.390 What is the germination requirement for solanaceous vegetables?

Ans. 70%

Q.391 Pollination behavior of bhendi is ________

Ans. Often-cross pollination

Q.392 Cross pollination in bhendi is due to the ________ nature of flower

Ans. Protogynous

Q.393 Isolation distance requirement for bhendi is ________

Ans. Foundation seed - 400 m
Certified seed - 200 m

Q.394 The immature and ill-filled seeds in bhendi are removed by________ technique.

Ans. Water floatation

Q.395 The objectionable weed species in bhendi seed production field is ______

Ans. Wild bhendi (*Abelmoschus moschata*)

Q.396 Bhendi seeds mature ______days after anthesis.

Ans. 30

Q.397 Physiological maturity symptoms of bhendi seed are ___________

Ans. - Drying of pods
- Changing of pod colour into straw yellow
- Development of hairline cracks along the ridges

Q.398 Delayed harvesting leads to __________ coloured seeds in bhendi

Ans. Off

Q.399 Seed recovery in bhendi is __________

Ans. 50% of total dry weight of fruit

Q.400 Which method of seed extraction is followed in bhendi?

Ans. Dry extraction

Q.401 Bhendi seed shows ___________dormancy

Ans. Hard seed coat

Q.402 Black coloured seed records ___________ germination than green colour seeds in bhendi.

Ans. Reduced

Q.403 Germination requirement for bhendi seed is ________

Ans. 65%

Q.404 Onion is highly __________pollinated crop

Ans. Cross

Q.405 Mechanisms favouring cross pollination in onion are________

Ans. Protoandrous
Nectar secretion (attracts insects)

Q.406 What is the name of the onion inflorescence?

Ans. Umbel

Q.407 Onion is a ________plant and acts as __________plant for seed production.

Ans. Long day; day-neutral

Q.408 How many seasons are required for seed production in onion?

Ans. Two seasons

Q.409 What are all the methods of seed production in onion?

Ans. Bulb to seed

Seed to seed

Q.410 Among two methods of seed production, which method seed has maximum genetic purity?

Ans. Bulb to seed

Q.411 The maximum seed yield is obtained in __________ method of seed production.

Ans. Seed to seed

Q.412 What is the temperature requirement for bolting in onion?

Ans. 10-15°C

Q.413 What is meant by bolting?

Ans. Initiation of flower stalk / flowering

Q.414 What are the types available in onion?

Ans. Bellary onion / large onion (*Allium cepa* var. *cepa*)
Country onion / small onion (*Allium cepa* var. *aggregatum*)

Q.415 What is the propagating material for bellary onion?

Ans. Sexual seed

Q.416 Country onion is propagated by using ____________ and ____________

Ans. Bulbs ; seeds

Q.417 In which type of country onion seeds are used as propagating material ?

Ans. Mutloor type

Q.418 What is the ideal temperature for storing the bulbs in bulb to seed method of seed production in onion?

Ans. 4-14°C

Q.419 What is the isolation distance for onion seed production?

Ans. Mother bulb production stage – 5 m
Seed production stage
Foundation seed – 1000 m
Certified seed - 500 m

Q.420 Bellary onion is a _______ crop and seedlings are raised in _____ bed nurseries.

Ans. Transplanted; raised

Q.421 What is the optimum depth of sowing for onion seeds?

Ans. 2 cm

Q.422 _______days old seedlings are transplanted in onion.

Ans. 30-35

Q.423 What is the optimum bulb size for the production of seed in bulb to seed method?

Ans. 40-50 g

Q.424 During replanting of bulbs in bulb to seed method the top 1/3rd portion is cut for the purpose of ________

Ans. Early germination and bolting

Q.425 The major pest in onion is ______

Ans. Thrips

Q.426 How many field inspections are required for onion seed crop?

Ans. Four

Q.427 What is the physiological maturity symptom for onion seed harvest?

Ans. Seed colour turns black

Q.428 How many harvests are required for onion?

Ans. 2 to 3

Q.429 What is the purpose of heaping the onion flower stalks after harvest ?

Ans. Completion of seed maturity and curing

Q.430 What is the germination requirement in onion seed?

Ans. 70%

Q.431 Hybrid seeds are produced by following the __________ and __________ methods in onion.

Ans. Cytoplasmic genic male sterility and self-incompatibility

Q.432 What is the planting ratio for onion hybrid seed production?

Ans. 4:1 (female : male)

Q.433 Most of the cucurbits are __________ plants in which both male and female flowers borne in the same plant at different positions

Ans. Monoecious

Q.434 Cucurbits are highly __________ pollinated crops.

Ans. Cross

Q.435 In a cucurbit vine which flower opens first, male or female?

Ans. Male flowers

Q.436 In which cucurbit crop the flower opening takes place in the evening hours?

Ans. Ridge gourd, bottle gourd

Q.437 Long photoperiod and high temperature induces more __________ flowers in cucurbits.

Ans. Male

Q.438 Chemical used to induce the femaleness in cucurbits is ___________

Ans. Etherel (ethephon) @ 200 ppm spraying at two leaf stage

Q.439 In cucurbits the ___________ is used for getting maximum plant population.

Ans. Pre-germinated seed

Q.440 What is the orientation of seed sowing in cucurbits?

Ans. Micropylar end upward

Q.441 What is the isolation distance requirement for cucurbits?

Ans. Foundation seed – 1000 m
Certified seed - 500 m

Q.442 What are the methods used for hybrid seed production in cucurbits?

Ans. Picking of male flowers and artificial pollination
Picking of male flowers and insect pollination
Use of gynoecious lines

Q.443 Seedless watermelon is developed from cross between ___________ and ___________ lines

Ans. Diploid; tetraploid

Q.444 What is the method of hybrid seed production followed in seedless watermelon?

Ans. Triploid

Q.445 What is the chemical used for the production of staminate flowers during hybrid seed production of cucurbits?

Ans. Gibberellic acid @ 100 ppm or silver nitrate @ 600 ppm

Q.446 What is the objectionable weed species for cucurbits?

Ans. Wild cucurbits

Q.447 What are all the seed borne viral diseases that cause severe damage to the cucurbit seed crop?

Ans. Cucumber mosaic virus, watermelon mosaic virus

Q.448 Indicate the different physiological maturity symptoms for different cucurbits.

Ans.

Ridge gourd	- fruit colour changes green to brown
Bitter gourd	- fruit colour changes green to yellow / orange
Bottle gourd	- fruit colour changes green to brown
Snake gourd	- fruit colour changes green to yellow / orange
Ash gourd	- fruit stalk becomes dry, ash coating on the fruit pericarp
Sponge gourd	- fruit colour changes green to brown
Water melon	- drying of tendrils at the point of attachment to the vine, ruffled dull or dead sound on thumping of fruits
Musk melon	- rind colour changes green to yellow / cream
Squash	- fruit colour changes green to deep golden yellow colour
Pumpkin	- fruit colour changes green to deep golden yellow colour

Q.449 What are all the cucurbit seeds extracted by wet extraction?

Ans. Snake gourd, bitter gourd, ash gourd, pumpkin, water melon, musk melon, cucumber, squashes.

Q.450 Dry seed extraction is practiced in __________cucurbits.

Ans. Ridge gourd, sponge gourd, bottle gourd

Q.451 Give few crop examples for wet seed extraction method.

Ans. Tomato, brinjal, bittergourd, pumpkin, snakegourd, melons, squashes.

Q.452 What are all the wet seed extraction methods used in cucurbits?

Ans. Fermentation method

Acid method

Mechanical method

Q.453 Which acid is used for cucurbits seed extraction?

Ans. Hydrochloric acid

Q.454 What is the germination percentage requirement for cucurbit seeds?

Ans. 60%

Q.455 In which crops over wintering (chilling) / vernalization is needed for shifting from vegetative to reproductive phase during seed production?

Ans. Cabbage, cauliflower, beets, carrots, turnips etc.

Q.456 What are the methods of seed production in temperate vegetable crops?

Ans.

Cabbage	- Head to seed, seed to seed
Cauliflower	- Curd to seed, seed to seed
Broccoli	- Head to seed, seed to seed
Knol-khol	- Head to seed, seed to seed
Carrot	- Root to seed, seed to seed
Radish	- Root to seed, seed to seed
Beet root	- Root to seed, seed to seed
Turnip	- Root to seed, seed to seed

Q.457 Where is head intact method followed?

Ans. Cabbage seed production

Q.458 Where is stump intact method of seed production followed?

Ans. Cabbage

Q.459 In cabbage __________method is mostly followed for nucleus seed production.

Ans. Head to seed

Q.460 What is the isolation distance for cruciferous vegetable crops?

Ans. Foundation seed - 1600 m

Certified seed - 1000 m

Q.461 What is the isolation distance for temperate root vegetable crops?

Ans. Foundation seed – 1600 m
Certified seed - 800 m

Q.462 Name the crops with highest isolation distance ?

Ans. Cabbage, beet

Q.463 Pollination behaviour of cruciferous vegetable crops is __________.

Ans. Cross pollination

Q.464 What is the mechanism favouring the cross pollination in cruciferous vegetables?

Ans. Self-incompatibility

Q.465 What is the main pollinating agent in cruciferous vegetable crops?

Ans. Honey bees

Q.466 The methods used for hybrid seed production in cruciferous vegetables are__________

Ans. Self-incompatibility and insect pollination
Cytoplasmic genic male sterility
Genetic male sterility

Q.467 Which hybrid seed production method is commercially exploited in crucifers vegetables?

Ans. Self-incompatibility and insect pollination

Q.468 What are the two types of self-incompatibility?

Ans. 1. Sporophytic self-incompatibility
2. Gametophytic self-incompatibility

Q.469 __________ self-incompatibility method is mostly used for hybrid seed production of cruciferous crops?

Ans. Sporophytic

Q.470 Steckling is used for planting in ___________crop seed production.

Ans. Cruciferous

Q.471 What is the planting ratio for hybrid seed production of crucifers vegetables?

Ans. 1:1 (female : male)

Q.472 Inflorescene of coriander and carrot is called as ________

Ans. Umbel

Q.473 The cross pollination in carrot is due to ____________nature

Ans. Protandrous

Q.474 Where do you find blonded seed?

Ans. Peas (due to cloudy weather)

Q.475 The main pollinating agent in carrot is __________

Ans. Honey bees

Q.476 Cytoplasmic male sterility is used in production of ________ hybrids

Ans. Flower crop

Q.477 Hybrid seed production method used in marigold is ________________

Ans. Hand emasculation and pollination

Q.478 Two main systems of MS lines in marigold hybrid seed production are ________and ________

Ans. Apetalous (stamenless); true double form

Q.479 The commercial hybrid seed production method in petunia is ________

Ans. Hand emasculation and pollination

Q.480 What is meant by plug mix?

Ans. Coir pith used for raising the flower crop seedlings

Q.481 Coconut is highly ________pollinated crop which requires ________ m isolation.

Ans. Cross; 3000

Q.482 The sowing orientation for coconut seed is ________

Ans. Vertical (stalk end upward)

Q.483 The age of the mother palm for selection of seednut in coconut is inbetween ______and ________years.

Ans. 25 and 40

Q.484 The seednuts in arecanut should be sown in ________ orientation

Ans. Vertical (stalk end upward)

Q.485 ________of areca seednuts results in early germination.

Ans. Dehusking

Q.486 Papaya is propagated by ________

Ans. Sexual seed

Q.487 The abortion of one ovule leads to the formation of single seeded fruit is called as _____in coffee.

Ans. Pea berry

Q.488 In ________plant the sex can be known only after six years

Ans. Nutmeg

Q.489 Bud grafting technique is followed in ________crop for its propagation.

Ans. Tea

Q.490 The name of the clove seed is ________

Ans. Mother of clove

Q.491 The part used for edible purpose in clove is called________

Ans. Unopened flower bud

Q.492 The mucilaginous layer on cardamom seed can be removed by _______treatment

Ans. 25% nitric acid

Q.493 What are the propagating materials for potato?

Ans. Seed tubers
True seeds

Q.494 What are minitubers?

Ans. Minitubers are potato tubers of 7-8 weeks old of 20 mm size and 4-5 gm

Q.495 What are microtubers?

Ans. Potato tubers raised in tissue culture labs of 1-2 cm

Q.496 What is meant by TPS?

Ans. True potato seed

Q.497 What is the method of hybrid seed production in potato?

Ans. Emasculation and dusting

Q.498 In which crop is dehaulming inspection is done ?

Ans. Potato

Q.499 Name a designated disease of potato ?

Ans. Brown rot

Q.500 Dehulming in seed potato crop is practiced to check _____________

Ans. Transmission of virus by aphids

Q.501 The physiological maturity is indicated by higher ______and ___________.

Ans. Dry matter; lower seed moisture content

Q.502 What is the physical purity percentage for maximum number of crops?

Ans. 98%

Q.503 Which crop requires minimum physical purity percentage and how much?

Ans. Daraf grass (*Pennisetum pedicellatum*)

Guniea grass (*Panicum maximum*) 80%

Q.504 What is the maximum level of physical purity requirement and in which crop seed?

Ans. Bhendi, 99%

Q.505 What is non pollen restorer line?

Ans. B line or Maintainer line

Q.506 What is pollen restorer line

Ans. R line or Restorer line

Q.507 What are bulbils?

Ans. Bulbils are apomicts where flowerbuds directly develop into young plants.

Q.508 Give an example for bulbils?

Ans. Agave

Q.509 Name a crop where the same crop should not have been grown for previous five seasons?

Ans. Buffel grass, Daraf grass, Marvel grass

Q.510 What is the prescribed number of counts needed for seed field more than 25 acre?

Ans. Nine

Q.511 What is the prescribed number of counts necessary for fields less than 5 acre?

Ans. Five

Q.512 What is the number of plants to be counted per count in case of wheat?

Ans. 1000

Q.513 What is the objectionable weed in mustard?

Ans. Mexican poppy (*Argemone mexicana*)

Q.514 What is the objectionable weed in sunflower?

Ans. Wild sunflower, Orabanche

Q.515 What is the minimum number of nodes on tapioca sett as per certification standard ?

Ans. Five

Q.516 What is the objectionable weed in lettuce?

Ans. Wild lettuce

Q.517 What are the inseparable other crops in wheat?

Ans. Barley, gram, oats, triticale

□□□

Chapter 6

Post Harvest Handling and Storage

Q.1 What is the cheapest seed drying method?

Ans. Sun drying or natural drying

Q.2 The machine used for drying seed is __________

Ans. Seed drier

Q.3 What is the optimum drying temperature for seed ?

Ans. 43°C

Q.4 The time allowed for attaining equilibrium moisture content during the drying process is called as ___________

Ans. Tempering period

Q.5 Specified quantity of seed, which is physically identifiable and homogenous is known as ________

Ans. Seed lot

Q.6 What is the name of the process in which a heterogeneous seedlot is made into homogenous one?

Ans. Seed processing

Q.7 What is the name of rule followed for blending seeds?

Ans. Pearson Square Rule

Q.8 What is meant by Dockage?

Ans. Impurities in the seed.

Q.9 What is meant by dunnage?

Ans. Tarpaulin or wooden or iron pallets used for staking seed bags in storage.

Q.10 What is meant by pre-cleaning?

Ans. Method of preparing the seeds for basic seed cleaning. Eg. Scalping, debearding, hulling, shelling etc.

Q.11 The machine used for pre-cleaning seed is called ________

Ans. Pre-cleaner

Q.12 Examples for pre-cleaner are____________

Ans. Scalper, debearder, huller, sheller, ginn, decorticators etc.

Q.13 Indicate the different pre-cleaners and their functions with examples.

Ans.

Machine	Usage	Crop
Debearder	Removal of appendages	Barley / Carrot
Decorticator	Shelling of pods	Groundnut
Sheller	Shelling of husk	Maize
Ginn	Removal of fuzz from seed	Cotton
Tomato pulper	Pulping of fruits	Tomato
Dewinger	Removal of wings	Pine

Q.14 Main cleaning operation during seed processing is referred as __________

Ans. Basic cleaning

Q.15 What is meant by scalping?

Ans. Removal of particles larger than seed

Q.16 What is the use of defuzzing machine?

Ans. It is used to remove hairs present on the seed eg. Tomato, carrot, sugarcane

Q.17 The machine used for basic cleaning is ________

Ans. Air screen cleaner cum grader

Q.18 What is meant by aspiration?

Ans. Removal of light seeds and chaffy material from the seed mass using negative air pressure or suction.

Q.19 The processing principle involved in air screen cleaner cum grader is ________

Ans. Seed size and weight

Q.20 Name the three screens available in seed grader.

Ans. Top screen, Middle screen and Bottom screen

Q.21 Seed quality is further upgraded by the use of ________ machines.

Ans. Up-grading

Q.22 Specify the name of few upgrading machines and their working principles.

Ans.

Machines	Principles
Roll mill	Seed shape and surface texture
Magnetic separator	Seed surface texture and stickiness
Electrostatic separator	Electrical property
Disc separator	Seed length
Spiral separator	Seed shape or the degree of its ability to roll
Specific gravity separator	Seed density or specific gravity
Inclined drapper	Shape and surface texture
Indented cylinder separator	Seed length
Vibratory separator	Seed shape, surface texture
Polisher	Seed lusture

Q.23 Upgradation of sunflower seed is done by ________ machine

Ans. Specific gravity separator

Q.24 Can certified seeds be processed in all processing units?

Ans. No, they can be processed only in notified processing unit

Q.25 Where is float test conducted and by whom?

Ans. Float test is conducted during rice seed processing by Seed Certification Officer to assess the percentage of floaters

Q.26 Maintenance of germination and vigour of seeds from harvest until it is planted is termed as________

Ans. Seed storage

Q.27 What are the different types of storage?

Ans. Storage of commercial seeds
Storage of carry – over seeds
Storage of foundation seed stocks
Storage of germplasm seed

Q.28 The important environmental factors that are affecting seed storability are______

Ans. Temperature
Relative humidity
Moisture content

Q.29 EMC means ___________

Ans. Equilibrium Moisture Content

Q.30 What is the safe seed moisture content for storability in moisture pervious and impervious containers?

Ans. Moisture pervious containers – 10-12%
Moisture impervious containers – 5-8%

Q.31 What is the ideal room temperature for storage of seeds?

Ans. 25°C

Q.32 The safe relative humidity for seed storage is __________

Ans. 40%

Q.33 The thumb rule for seed storage was given by ___________

Ans. Harrington,1972

Q.34 For every decrease of 1% seed moisture the storage life of seed is ______and this rule is applicable in the seed moisture content range of ___

Ans. Doubled; 5 and 14%

Q.35 For every decrease of 5°C in storage temperature the storage life of seed is ______ and the rule is applicable within the range of _______

Ans. Doubled; 0° to 50°C

Q.36 The sum of RH in % and the temperature in °F of storage environment in Harrington's numerical thumb rule should be upto __________ for good seed storage but the contribution from temperature should not exceed__________

Ans. 100; 50°F

Q.37 The combination of temperature and RH in determining storage climate is known as ________

Ans. Numerical thumb rule

Q.38 Increase in concentration of the gases like_______will increase the seed storage life.

Ans. Carbon di-oxide (CO_2) and Nitrogen (N_2)

Q.39 The high concentration of_______gas decreases seed storability.

Ans. Oxygen (O_2)

Q.40 __________and ___________are the two important factors influencing the lifespan of seeds during storage.

Ans. Temperature; relative humidity

Q.41 What is the meaning of 'modified atmospheric seed storage'?

Ans. Regulation of temperature and relative humidity in storage container / godown

Q.42 What is the use of Grain Micrometer?

Ans. It is a device used to measure the temperature and relative humidity of an atmosphere

Q.43 Give examples for moisture and vapour pervious containers?

Ans. Jute bag, cloth bag

Q.44 Give examples for moisture impervious and vapour pervious containers?

Ans. Polyvenyl bags, polythene bags of less than 300 gauge thickness

Q.45 Examples for moisture and vapour proof containers are ___________

Ans. Polythene bags of more than 700 gauge thickness, aluminium foil pouches, rigid plastic containers and tins.

Q.46 Gas commonly used for modified atmospheric storage is ___________

Ans. CO_2

Q.47 What is the desiccant used for seed storage?

Ans. Silica gel

Q.48 What are all the fumigants used for seed storage?

Ans. Methyl bromide

Hydrogen cyanide

Ethylene dibromide

Carbon tetra chloride

Carbon disulphide

Napthalene

Q.49 Commonly used fumigant in seed storage godown is ________

Ans. Aluminium phosphide

Q.50 Which crop seed can be desiccated to below 4% and still remain viable ?

Ans. Sesame

Q.51 What is the storage moisture content for potato ?

Ans. Not specified

Q.52 Treating the seeds with protectants like fungicide or insecticide or bio-control agent or vigour improving agent is termed as __________

Ans. Seed treatment

Q.53 Types of seed treatment are __________

Ans. 1. Pre-sowing seed treatment
2. Post harvest or pre-storage seed treatment
3. Mid-storage seed treatment

Q.54 Specify the different types of pre-sowing seed treatments?

Ans.
- Seed hardening
- Seed fortification
- Seed infusion
- Seed priming
- Seed pelleting
- Seed coating
- Seed colouring
- Seed scarification
- Seed stratification
- Fluid drilling

Q.55 Mention the different types of pre-storage seed treatments?

Ans.
- Dry seed treatment
- Wet seed treatment/Slurry seed treatment
- Organic infusion
- Vapour phase technique

Q.56 The mid-storage seed treatments include__________

Ans. Soaking - drying seed treatment
Dipping - drying seed treatment
Spraying - drying seed treatment
Moisture equilibration seed treatment
Moist sand conditioning and drying
Halogenation

Q.57 Treatment given to seed to invigorate the quality is termed as______

Ans. Seed fortification or seed invigouration

Q.58 What are the chemicals used for seed fortification?

Ans. Macronutrients, micronutrients, vitamins, growth regulators etc.

Q.59 Seeds that are susceptible to soaking injury and seeds with internal seed borne diseases which need penetration of chemical to kill the pathogen is given with the pre-sowing seed treatment called ________

Ans. Seed infusion

Q.60 The solvents used for seed infusion are ________

Ans. Acetone, alcohol

Q.61 Coating the seed without altering its shape is termed as ________

Ans. Seed coating

Q.62 The size and shape of the seed is altered in the pre-sowing seed treatment called____________

Ans. Seed pelleting

Q.63 The chemicals used for seed pelleting are___________

Ans. Macronutrients, micronutrients, vitamins, growth regulators, pesticides etc.

Q.64 What are the filler materials used for seed pelleting?

Ans. Gypsum
Calcium Carbonate (Chalk powder)
Charcoal
Flyash
Woodash
Leaf powders (*Albizia, Prosopis, Pungamia, Vitex*)

Q.65 What are the adhesives used for seed pelleting?

Ans. Maida - 10%
Starch - 5%
Rice gruel
Revive - 2%
Sago gruel
Gum arabic

Q.66 Colouring the seeds with colouring agents is known as ________________

Ans. Seed colouring

Q.67 The chemicals or dyes used for seed colouring are ____________

Ans. Methyl red
Bromocresol green
Methylene blue
Rose Bengal

Q.68 Seed treatment given to seed for arid-sowing or drought regions is termed as_________

Ans. Seed hardening

Q.69 What are the chemicals used for seed hardening?

Ans. Potassium chloride (KCl), Sodium dihydrogen orthophosphate (NaH_2PO_4), Potassium dihydrogen orthophosphate (KH_2PO_4)

Q.70 The ratio of seed and solution for pulses seed hardening is ____________

Ans. 1:0.33

Q.71 The initiation of seed germination process is achieved by _______ treatment

Ans. Seed priming

Q.72 Soaking the seed in salt solutions is called __________ priming

Ans. Halo

Q.73 What is meant by solid matrix priming?

Ans. Seed priming using inert materials viz., sand, sawdust, peat etc.

Q.74 What is meant by bio-priming?

Ans. Use of biological substances or agents for priming the seed.

Q.75 IDS is mostly suitable for _________ crops

Ans. Tree (eg.Pine)

Q.76 Seed ________ refers to the destruction of surface borne organisms that have contaminated the seed surface but not infected the seed.

Ans. Disinfestation

Q.77 Aerated steam therapy is a seed treatment used in _________ crop to control ________ disease.

Ans. Sugarcane ; grassy shoot

Q.78 The insecticides used for seed treatment are__________

Ans. Sevin (200 mg/kg of seed) or Malathion D (0.06 mg/kg of seed)

Q.79 What is the new chemical compound used for seed treatment to protect the crop from sucking pests?

Ans. Imidacloprid

Q.80 What are all the fungicides used for seed treatment?

Ans. Thiram / Captan / Bavistin @ 2 g / kg of seed

Q.81 Name the seed treatment given during mid way of storage?

Ans. Hydration-dehydration

Q.82 The seed moisture content increase in hydration treatment is ______

Ans. 25-30%

Q.83 List out the chemicals used for hydration-dehydration seed treatment

Ans. Sodium phosphate, potassium phosphate, potassium di hydrogen ortho phosphate, sodium chloride, para hydroxy benzoic acid, potassium iodide, para amino benzoic acid.

Q.84 What is the important chemical used for mid storage correction in seed?

Ans. Potassium di hydrogen orthophosphate

Q.85 What is the concentration of chemicals used in mid-storage hydration-dehydration treatment?

Ans. 10^{-3} to 10^{-4} M

Q.86 What is the name of the vapour seed treatment?

Ans. Chlorination and iodination / Halogenation

Q.87 Chemicals used for vapour seed treatment?

Ans. Chlorine, iodine, bromine

Q.88 What is meant by vapour phase seed treatment?

Ans. Treating / impregnating the seeds with vapours of chlorine or iodine or bromine

Q.89 The other name of vapour phase seed treatment is______

Ans. Halogenation seed treatment

Q.90 The most commonly used halogen for seed treatment is__________

Ans. Calcium oxy chloride ($CaOCl_2$)

Q.91 The carrier material used for halogenation seed treatment is ___________

Ans. Calcium carbonate ($CaCO_3$)

Q.92 The two types of germplasm maintenance are __________

Ans. (i) *Insitu* conservation (ii) *Exsitu* conservation

Q.93 Maintenance of plant species within its natural environment is known as _____

Ans. *Insitu* conservation

Q.94 Examples for *insitu* conservation are __________

Ans. National Park, Biosphere Preserves

Q.95 Collection of germplasm species from different parts of world into a plant repository is termed as __________

Ans. *Exsitu* conservation

Q.96 An example for *exsitu* conservation is __________

Ans. Kew Garden, England

Q.97 Different types of *exsitu* conservation are __________

Ans. Active and base collections
Mass resources of composites
Field conservation
In-vitro conservation

Q.98 The species which are poor seed producers or species which cannot be stored under ordinary storage methods can be preserved by __________

Ans. *Invitro* conservation

Q.99 Plant materials under *invitro* method can be stored at low temperature of ____

Ans. -196^{0}C (Liquid Nitrogen (LN))

Q.100 ____________are some of the cryoprotectants used to avoid freezing injury to the cells of the tissues stored under liquid nitrogen.

Ans. Dimethyl sulfoxide (DMSO)
Poly ethylene glycol (PEG)

Q.101 ____________is the genome material used under advanced method of species conservation which is called __________

Ans. DNA; DNA Bank

Q.102 The first Asian DNA Bank is established at _________in India

Ans. Lucknow

Q.103 The DNA bank in India is the worlds ________bank

Ans. Second

Q.104 The seed bank located in Artic area is _________

Ans. Svalbard Global Seed Vault

Q.105 Svalbard Global Seed Vault is located at _______

Ans. Norwegian island of Spitsbergen near the town of Longyearbyen

Q.106 Primary funding agency for Svalbard Global Seed Vault is ____________

Ans. Bill and Melinda Gates Foundation

Q.107 High altitude storage (seed vault) in India is located at

Ans. Defence Institute of High Altitude Research in Chang-La, Jammu Kashmir

Q.108 When was the seed vault in India established ?

Ans. November, 2009

❏❏❏

Chapter 7

Seed Viability and Vigour

Q.1 **What is meant by seed viability in short?**
Ans. State of aliveness

Q.2 **What is meant by seed vigour in short?**
Ans. Degree of aliveness

Q.3 **What is the other name of ageing in seed?**
Ans. Deterioration

Q.4 **Specify the name of the enzymes where its activity is reduced during seed deterioration?**
Ans. Alpha-amylase, catalase, peroxidase, dehydrogenase, cytochrome oxidase

Q.5 **Which enzyme activity gets increased during seed ageing?**
Ans. Phytase, protease, phosphatase

Q.6 What is the phenolic compound excessively accumulated in aged rice seed?

Ans. Ferulic acid

Q.7 Mention the classes of seed based on its storability?

Ans. 1. Microbiotic seed
2. Mesobiotic seed
3. Macrobiotic seed

Q.8 Mention the classes of seed based on its chemical composition?

Ans. Cereals (carbohydrate)
Pulses (protein)
Oilseeds (fat)
Spices (aromatic acids)

Q.9 What is the name of the seeds based on its moisture content and storability?

Ans. Orthodox
Recalcitrant
Intermediate

Q.10 Which seed possess maximum viability period?

Ans. Lupin (*Lupinus alpinum*)

Q.11 What is the viability period of lupin seed?

Ans. 10,000 years

Q.12 Mention the name of the seed having the viability period of 1000 years?

Ans. Lotus (*Nelumbo nucifera*)

Q.13 What is the viability period of cassia seeds?

Ans. 100 years

Q.14 What is meant by orthodox seed?

Ans. Seeds having tolerance to desiccation and freezing sensitivity and long viability period.

Q.15 Give the term for desiccation and freezing sensitive seeds?
Ans. Recalcitrant seeds

Q.16 Give the examples for orthodox seed?
Ans. Seeds of cereals, pulses, oilseeds, fibres, vegetables etc.

Q.17 What is the viability period of recalcitrant seed?
Ans. Few days to few weeks or months

Q.18 Examples for recalcitrant crop seeds?
Ans. Some species of fruit crops, plantation crops, spice crops and some forest species

Q.19 Examples for recalcitrant fruit crop seeds?
Ans. Mango, mangosteen, avocado, jamun

Q.20 Examples for recalcitrant seeds in plantation crops?
Ans. Cocoa, rubber, coconut, arecanut

Q.21 Examples for recalcitrant seeds in spice crops?
Ans. Nutmeg, clove, cinnamon, curry leaf

Q.22 Cocoa seed is a desiccation ____________seed.
Ans. Sensitive

Q.23 Examples for recalcitrant forest seeds?
Ans. *Araucaria excelsa, Rhizophora* sp., *Dipterocarpus* spp., *Shorea* spp.

Q.24 Expand LSMC?
Ans. Lowest Safe Moisture Content

Q.25 What is meant by LSMC?
Ans. The safe moisture content below which desiccation damage occurs in recalcitrant seeds

Q.26 LSMC for mango is _________
Ans. 25-35% (varietal specific)

Q.27 What is the lowest safe moisture content requirement for cocoa?
Ans. 26%

Q.28 Give the LSMC for arecanut?
Ans. 32.8%

Q.29 What is the LSMC for *Shorea* sp.?
Ans. 30% (*S. robusta*)

Q.30 Which species requires the maximum level of LSMC?
Ans. *Vatica umbonata* (74%)

Q.31 What is the lowest safe storage temperature for recalcitrant seeds?
Ans. 20°C

Q.32 Indicate the LSMC for intermediate class of seeds?
Ans. 12-14%

Q.33 What is the level of low temperature tolerance of intermediate seeds?
Ans. 2-5°C

Q.34 Give an example for viviparous recalcitrant seed?
Ans. *Rhizophora* sp.

Q.35 Give an example of viviparous seed in cultivated crop?
Ans. Chow chow (*Sechium edule*)

Q.36 What is the other name of LSMC?
Ans. Critical moisture content

Q.37 Give expansion for LEA?
Ans. Late Embryogenic Abundant (LEA) proteins

Q.38 What is the role LEA protein?
Ans. It imparts desiccation tolerance to the recalcitrant seeds

Q.39 Specify the short term storage methods for recalcitrant seeds?

Ans. Moist or imbibed storage, partial dehydration, controlled atmospheric storage

Q.40 What is the long term storage method for recalcitrant seeds?

Ans. Cryopreservation and *in vitro* conservation of excised embryo

Q.41 What are all the growth hormones produced during seed germination?

Ans. Gibberellic acid, cytokinin, ethylene

Q.42 Name the hormone abudantly produced during the seed germination?

Ans. Gibberellic acid

Q.43 When does seed start ageing?

Ans. Immediately after detachment from mother plant

Q.44 What are all the basic causes of ageing in seeds?

Ans. Accumulation of toxic substances, destruction of enzymes and loss of membrane integrity

Q.45 What is the best method to assess the membrane integrity?

Ans. Measurement of electrical conductivity of seed leachates

Q.46 What are all the compounds released from aged seed while soaking in water?

Ans. Inorganic salts, amino acids, nucleotides, sugars, free fatty acids

Q.47 What is the name of the process of degradation of fat into free fatty acids in seeds?

Ans. Lipid auto-oxidation

Q.48 Degree of aliveness / aggressivity of seeds is known as __________

Ans. Vigour

Q.49 Indicate different vigour levels in seed?

Ans. Low vigour seeds
Medium vigour seeds
High vigour seeds

Q.50 Expand the abbreviation VI?

Ans. Vigour index

Q.51 How vigour index is calculated?

Ans. Vigour index = Germination (%) x seedling length (cm) or Germination (%) x Dry matter production (mg/g).

Q.52 Who has given the formula for vigour index?

Ans. Abdul-Baki and Anderson (1971)

Q.53 What is the name of the vigour test in which the seed leachate is measured?

Ans. Electrical Conductivity test (EC test)

Q.54 Which seed leaks more electrolytes in imbibing water?

Ans. Low vigour seeds

Q.55 What is meant by GADA test in seed vigour?

Ans. Glutamic Acid Decarboxylase Activity test

Q.56 What are all the important stress tests for seed vigour assessment?

Ans. Accelerated ageing test
Cold test
Brick gravel test
Osmotic test

Q.57 What is meant by AAT or AA test?

Ans. Accelerated Ageing test

Q.58 Who developed the accelerated ageing test for seed vigour measurement?

Ans. Delouche (1965)

Q.59 What is the temperature and relative humidity level of accelerated ageing test?

Ans. 40-45°C, 100% RH

Q.60 In which crop the cold test is followed for seed vigour assessment?

Ans. Maize

Q.61 Which crop vigor is estimated using cool germination test?

Ans. Cotton

Q.62 Who found the cold test for seed vigour status assessment?

Ans. Isley (1950)

Q.63 Who developed the brick gravel test for seed vigour measurement?

Ans. Hiltner and Ihssen (1911).

Q.64 What are all the osmatica used for vigour test?

Ans. Mannitol, polyethylene glycol (PEG), sucrose, glycerol.

Q.65 What is meant by SGR in seed vigour test?

Ans. Seedling Growth Rate.

Q.66 Who first described the vigour test, speed of germination?

Ans. Kotowski (1926)

Q.67 Who has given the formula for calculating speed of germination?

Ans. Maguire (1962).

Q.68 What is the main factor for sudden decrease in seed vigour?

Ans. Mechanical injury.

Q.69 What is the other name of brick-grit test?

Ans. Hiltner's test

Q.70 Who gave exhaustion test?

Ans. Germ (1960)

Q.71 Mention New Seed Vigour test using computer.

Ans. SVIS (Seed Vigour Imaging System)

Q.72 'SSAA' is expanded as ___________

Ans. Saturated Salt Accelerated Ageing test

Q.73 'SSAA' test is used to assess _________

Ans. Seed Vigour

Q.74 Who developed SSAA test ?

Ans. Jianhua and McDonald (1996)

□□□

Chapter 8

Seed Testing

Q.1 **Who is known as the father of seed testing?**

Ans. Prof. Fredrich Nobbe

Q.2 **Where was the first seed testing lab located?**

Ans. Thrandt, Saxony, Germany

Q.3 **When was the first seed testing lab established?**

Ans. 1869

Q.4 **Science of evaluating the planting value of seed before it is sown is called ___**

Ans. Seed testing

Q.5 **Which crop has the largest lot size?**

Ans. Maize (40000kg)

Q.6 **A small portion of seed taken from different representative parts of the lot is called _____**

Ans. Primary sample

Q.7 How many numbers of primary samples should a seed certification officer take?

Ans. Three

Q.8 What is the name of the mixture of all primary samples?

Ans. Composite sample

Q.9 Sample submitted to the seed testing laboratory is termed as __________

Ans. Submitted sample

Q.10 What is the minimum number of seeds in a submitted sample?

Ans. 2500

Q.11 What is the submitted sample size for moisture estimation in case of grinding ?

Ans. 100 g

Q.12 In which crop the submitted sample is provided in numbers ?

Ans. Aggregate onion, chow-chow

Q.13 What is the submitted sample size for pelleted seeds ?

Ans. 7500 Nos.

Q.14 Sample used for quality test is called as ________?

Ans. Working sample

Q.15 What is the sample stored in seed testing lab after testing known as?

Ans. Guard sample

Q.16 What is referee sample?

Ans. For every 5000 submitted samples tested by the state seed testing lab, one sample is sent to the central seed testing lab which acts as the referee lab, for verifying the authenticity of testing.

Q.17 What is 'H' value?

Ans. H value provides the heterogeneity value of a seed lot.

Q.18 Device used for seed sampling is called ________?

Ans. Triers or probes

Q.19 Indicator siliga gel is mixed with __________ to indicate color change upon absorbing moisture

Ans. Lithium chloride

Q.20 The machines used for dividing the seed samples is known as ________

Ans. Dividers

Q.21 What is the working principle of Gamet divider?

Ans. Centrifugal force

Q.22 Which divider is most precise?

Ans. Gamet divider

Q.23 Name the divider that works under the principle of gravitational force

Ans. Boerner Divider

Soil Divider

Riffle Type Divider

Q.24 How will you mix and divide fuzzy cotton seed?

Ans. Manually

Q.25 How will you mix and divide mustard seed?

Ans. Modified halving method

Q.26 What is the other name of conical divider?

Ans. Boerner Divider

Q.27 How should the sample for assessment of moisture submitted?

Ans. In sealed plastic bag

Q.28 What are all the components of purity analysis?

Ans. Pure seed, inert matter

Q.29 What is the equipment used for purity analysis?

Ans. Purity work board

Q.30 Physical purity of the seed can be improved by ________

Ans. Seed processing

Q.31 In which crop decoated seeds are considered as inert matter?

Ans. Pulses

Q.32 In which crop pieces of mericarps are considered as pure seed?

Ans. Coriander

Q.33 Can nematode gall be considered as pure seed?

Ans. No

Q.34 Which rule states that pieces of seed more than half the original seed are considered as pure seeds?

Ans. Fifty percent seed rule

Q.35 Which crop has the highest acceptable per cent of inert matter?

Ans. Buffel grass (20%)

Q.36 Which crop has the highest physical purity?

Ans. Bhendi (99%)

Q.37 Which crop has the lowest physical purity?

Ans. Buffel grass, Guniea grass (80%)

Q.38 In which crop there is no specific standard for purity percentage?

Ans. Aggregate onion, potato

Q.39 In what crop, huskless seeds are determined during purity analysis?

Ans. Rice

Q.40 Expand 'PLS'________

Ans. Pure Live Seed

Q.41 PLS can be calculated by ________

Ans. $PLS(\%) = \frac{Purity(\%) \times Germination(\%)}{100}$

Q.42 Name the accurate method for seed moisture estimation.

Ans. Hot air oven method

Q.43 What is the temperature requirement for "low constant temperature method" of moisture estimation?

Ans. 103°C

Q.44 Menticn the temperature requirement for "high constant temperature method" of moisture estimation?

Ans. 130°C

Q.45 What is the predrying moisture content for wheat ?

Ans. 17%

Q.46 What is the predrying moisture content for castor ?

Ans. 13%

Q.47 State of aliveness of seed is called ________?

Ans. Viability

Q.48 What is the role of nomograph ?

Ans. To predict the seed viability period

Q.49 Who invented viability nomograph?

Ans. E.H.Roberts

Q.50 What is the name of the rapid test used for seed viability assessment?

Ans. Tetrazolium test

Q.51 Mention the other names of Tetrazolium test.

Ans. Quick viability test
2,3,5-Triphenyl Tetrazolium test
TTZ test
TTC test
Prediction test
Topographical Tetrazolium test

Q.52 Who gave the tetrazolium test for seed viability?

Ans. Prof. George Lakon

Q.53 What is the name of the compound formed during tetrazolium test?

Ans. Formazan

Q.54 What is used to extract triphenyl formazan?

Ans. Alcohol or methyl cellosolve

Q.55 What is the name of the enzyme involved during tetrazolium test?

Ans. Dehydrogenase enzyme

Q.56 What are the names of the viability tests for tree seeds?

Ans. Indigo Carmine (IC) test
Radiographic (X-ray contrast) test
Embryo Excision (EE) test

Q.57 What is meant by 'Vitascope'?

Ans. It is a device used to conduct the tetrazolium test

Q.58 How many seeds are used for conducting germination test as per ISTA norms?

Ans. 400 seeds

Q.59 Expand 'TP'

Ans. Top of paper

Q.60 'BP' means______

Ans. Between paper

Q.61 Expand 'PP' method

Ans. Pleated Paper method

Q.62 Substratum used for conducting germination test is ___________

Ans. Paper
Sand
Soil

Q.63 Which is the best germination test method for light requiring seeds?

Ans. Top of paper

Q.64 Optimum temperature requirement for seed germination is ____

Ans. 25°C

Q.65 Light requirement for conducting germination test is ________

Ans. 1250 Lux

Q.66 Equipment used to determine the temperature requirement for seed germination ________

Ans. Thermogradient table

Q.67 Instrument used to measure humidity in germination room is called________

Ans. Hygrometer

Q.68 What is the use of thermohygrograph?

Ans. It is a device used to measure the temperature and relative humidity.

Q.69 What is Rodewald apparatus used for?

Ans. Germination

Q.70 What is meant by "Copenhagen tank"?

Ans. Device used to test the germination of vegetable and flower crop seeds.

Q.71 The crop seed that requires minimum number of days for its final evaluation during laboratory germination test is ___________

Ans. Crop: Jute - 5 days required

Q.72 The crop seed that requires maximum number of days for its final evaluation during laboratory germination test is ___________

Ans. Spinach (35 days)

Q.73 Which crop has the lowest acceptable germination percentage?

Ans. Daraf grass (15%)

Q.74 In which crop no specific germination standards have been provided ?

Ans. Aggregate onion, potato

Q.75 What crop has to be shelled before placing for germination ?

Ans. Groundnut

Q.76 Which crop seed require light for germination ?

Ans. Ashgourd

Q.77 Which crop requires pre-washing prior to placing for germination ?

Ans. Spinach beet

Q.78 Which grass seed requires acid scarification for germination ?

Ans. Dinanath grass

Q.79 The least number of days required for first count during germination test evaluation is ________

Ans. 3 days

Q.80 What are all the chemicals used for breaking seed dormancy during germination test?

Ans. Potassium nitrate, gibberellic acid

Q.81 What is the chilling temperature requirement to conduct germination test?

Ans. 5-10°C

Q.82 ____________ is the pre-drying temperature required for conducting the laboratory germination test.

Ans. 40°C

Q.83 What is meant by 'Grey Process" of seed treatment?

Ans. Commercial method of seed treatment in an electromagnetic field

Q.84 What is '4D's of abnormal seedlings in germination test evaluation ?

Ans. Damaged seedlings
Deformed seedlings
Decayed seedlings
Deranged seedlings

Q.85 'Knee bend' is common in ______ crop seedlings during germination test.

Ans. Onion

Q.86 What is the rule that states that absence of one cotyledon can be considered as normal seedling?

Ans. Fifty percent seedling rule

Q.87 Tests to identify mechanical injury in seeds are __________

Ans. Indoxyl Acetate test, Ferric chloride test, Fast green test

Q.88 Name a crop where hard seed is included in germination per cent ?

Ans. Blackgram

Q.89 What is the important enzyme present in the germinating seed?

Ans. Alpha-amylase

Q.90 When would you go for retesting in germination?

Ans. When the difference between maximum and minimum values in germination test exceeds the tolerance limit.

Q.91 Test to identify mechanical injury in corn seeds is called as ____________

Ans. Ferric chloride test

Q.92 Sodium hypochlorite test is used to identify seed coat injury in ________seed

Ans. Soybean

Q.93 Cutting test is followed in _________ seed.

Ans. Pine

Q.94 What are all the methods used for assessing the genetic purity of the crop varieties?

Ans. Physical tests
Biochemical tests
Morphological tests
Molecular tests

Q.95 Expand ODV ?

Ans. Other Distinguishable Variety

Q.96 Give an example for physical test for assessing genetic purity

Ans. Other Distinguishable Variety test (ODV test)

Q.97 Give few examples for biochemical tests?

Ans. KOH bleach test
NaOH test
Phenol test
Peroxidase test
Electrophoresis

Q.98 What is the enzyme involved in phenol test?

Ans. Tyrosinase

Q.99 Examples for morphological test are_________

Ans. Coleoptile colour
DDT resistance
Field plot or grow out test (GOT)

Q.100 What is the principle of electrophoresis?

Ans. Migration of a charged particle under the influence of an electric field.

Q.101 What are the factors that affect the movement of the particles in the electrophoresis technique?

Ans. Electric charge
Size of the particle

Q.102 What are the molecules separated by electrophoresis for varietal identification?

Ans. Proteins
Isozymes

Q.103 What are the two approaches of electrophoresis?

Ans. Direct approach – separation of seed proteins
Indirect approach – separation of isozymes

Q.104 Name the wheat seed portein used for varietal identification by electrophoresis

Ans. Gliadin

Q.105 Name the crop in which hordein protein is used for varietal identification by electrophoresis.

Ans. Barley

Q.106 Zein protein is used in ____crop for varietal identification by electrophoresis.

Ans. Maize

Q.107 What is the name of the seed protein used for varietal identification by electrophoresis in chickpea?

Ans. Lectin

Q.108 What is meant by PAGE?

Ans. Poly Acrylamide Gel Electrophoresis

Q.109 What is meant by 'IEF' in Electrophoresis?

Ans. Iso Electric Focusing

Q.110 Expand the name 'UTIEF'

Ans. Ultra Thin Isoelectric Focusing

Q.111 What is the important stain used for staining the gel during electrophoresis ?

Ans. Comassie Brilliant Blue

Q.112 What is meant by Rm value in electrophoresis?

Ans.

$$\text{Relative mobility (Rm)} = \frac{\text{distance travelled by the band}}{\text{distance travelled by the tracking dye}}$$

Q.113 Varietal purity are assessed by electrophoresis through the preparation of ______

Ans. Electrophorogram

Q.114 What are the different molecular methods used for the genetic purity test?

Ans. RAPD (Random Amplified Polymorphic DNA)
RFLP (Restriction Fragment Length Polymorphism)
PCR (Polymerase Chain Reaction)
AFLP (Amplified Fragment Length Polymorphism)
Micro-satellites - SSR (Simple Sequence Repeats)
STR (Short Tandem Repeats)
ISSR (Inter-Simple Sequence Repeats)
STMS (Sequence Tagged Microsatellites Sites)
VNTR (Variable Number Tandem Repeats)
SNP (Single Nuclotide Polymorphism)

Q.115 Which molecule is used for molecular varietal purity test?

Ans. DNA

Q.116 What is the important technique that comes under probe based DNA profiling?

Ans. RFLP (Restriction Fragment Length Polymorphism)

Q.117 What molecular techniques are called as second generation techniques?

Ans. RAPD (Random Amplified Polymorphic DNA)
RFLP (Restriction Fragment Length Polymorphism)
AFLP (Amplified Fragment Length Polymorphism)
STMS (Sequence Tagged Microsatellites Sites)

Q.118 What molecular techniques are called as third generation techniques?

Ans. SNP (Single Nuclotide Polymorphism)

Q.119 'GM' crops are called as________

Ans. Genetically Modified crops

Q.120 The technology involved to produce GM crop is called as ____________

Ans. Genetic Engineering or Genetic Modification

Q.121 Transfer of foreign gene into the crop plants lead to ________

Ans. Transgenic plants or GM crop plants

Q.122 ________are few examples of transgenic crop plants.

Ans. *Bt* - cotton, *Bt* - maize, Roundup Ready Soybean, Golden rice, *Bt* - brinjal.

Q.123 What is the commercial transgenic technology used in cotton?

Ans. *Bt* - technology or *Bt* - genes

Q.124 *Bt* - means__________

Ans. *Bacillus thuringiensis*

Q.125 Which *Bt* - gene is most successful in cotton transgenic crops?

Ans. Cry 1 AC

Q.126 Cotton *Bt* - technology is resistant against the pest called ________

Ans. Bollworms

Q.127 The Cry 1 AC produces a ___________ called _______ which have insecticide properties

Ans. Crystalline protein; Endotoxin

Q.128 The *Bt* - genes are inserted into the cotton plants through __________bacteria

Ans. *Agrobacterium tumifacians*

Q.129 The development of resistance to the Bt genes in cotton plants can be prevented by growing the non-*Bt* - plants known as ________

Ans. Refugee plants

Q.130 The presence of *Bt* - gene in the cotton seeds can be detected even in the commercial selling points by the simple technology called ______

Ans. *Bt* - Express or *Bt* - strip

Q.131 *Bt* - Express strip for cotton was developed in India by the institute ____________

Ans. Central Institute for Cotton Research (CICR), Nagpur

Q.132 Genetic engineering technology that kills seed is called as _________

Ans. Technology Protection System / Terminator Technology

Q.133 What is meant by "GURT"?

Ans. Genetic Use Regulation Technology, Terminator Technology

□□□

Chapter 9

Seed Health

Q.1 Pathogen incidence in the seed is known as ?

Ans. Seed infection

Q.2 Insect incidence in the seed is known as ?

Ans. Seed infestation.

Q.3 What is the term for disease causing pathogen transmission by seed?

Ans. Seed borne

Q.4 Give few examples for embryo transmitted pathogens in seed?

Ans. Loose smut of wheat
Loose smut of barley
Anthracnose leaf spot in cotton
Bean common mosaic virus
Tobacco ring spot virus
Barley stripped mosaic virus

Q.5. Seed borne diseases specific to seed certification is known as ____________.

Ans. Designated disease

Q.6 Barley loose smut (designated) disease is caused by __________

Ans. *Ustilago nuda*

Q.7 Oats designated disease, Loose smut is caused by __________

Ans. *Ustilago avenae*

Q.8 *Ustilago triticii* pathogen causes __________disease in wheat and it is a __________disease.

Ans. Loose smut, designated

Q.9 Karnal bunt in wheat caused by ________

Ans. *Tilletia indica* (*Neovossia indica*)

Q.10 What is the seed borne pathogen in rice?

Ans. *Helminthosporium oryzae* (or) *Diplodia oryzae*

Q.11 What is the bunt disease causing organism in rice?

Ans. *Neovossia horrida*

Q.12 What is to be sprayed to prevent khaira disease in rice?

Ans. 0.5% Zinc sulphate and 0.25% lime

Q.13 Pathogen that causes ergot disease in pearlmillet is__________and it is a ______disease.

Ans. *Claviceps fusiformi*, designated

Q.14 The designated disease, smut in pearlmillet is caused by___________

Ans. *Tolyposporium penicillariae*

Q.15 *Sclerospora graminicola* pathogen causes ________ disease (designated disease) in pearlmillet.

Ans. Green ear

Q.16 Anthracnose disease in sorghum is caused by __________

Ans. *Colletotrichum graminicola*

Q.17 The designated disease, grain smut in sorghum is caused by_________

Ans. *Sphacelotheca sorghi*

Q.18 What is the causal organism of sorghum head smut (designated) disease?

Ans. *Sphacelotheca veiliana*

Q.19 In maize, southern leaf blight is caused by

Ans. *Drechslera veiliana*

Q.20 Black kernel of maize is caused by

Ans. *Botrydiplodia theobromae*

Q.21 What is the name of the disease caused by *Fusarium moniliforme* in maize?

Ans. Cob rot

Q.22 The infection point of *Diplodia oryzae* and *D. zeae* in seed is _________

Ans. Endosperm

Q.23 What is the disease causing organism for maize stalk rot?

Ans. *Diplodia zeae*

Q.24 What are all the designated diseases in cowpea?

Ans. Ashy stem blight - *Macrophomina phaseoli*

Anthracnose - *Colletotrichum lindemuthianum*

Ascochyta blight - *Ascochyta* sp.

Q.25 What is the designated disease in green gram?

Ans. Halo blight (*Pseudomonas phaseolicola*)

Q.26 Soybean seed discolouration (Purple seed stain) is caused by

Ans. *Cercospora kikuchii*

Q.27 Anthracnose leaf spot in soybean is caused by ________

Ans. *Colletotrichum dematinum*

Q.28 Anthracnose wilt in cotton is caused by ________

Ans. *Fusarium oxysporum* f. sp. *vasinfectum*

Q.29 What is the name of the seed borne causal organism for anthracnose disease in cotton?

Ans. *Colletotrichum gossippi*

Q.30 What is the designated disease in sesame?

Ans. Leaf spot (*Cercospora sesami*)

Q.31 What is the designated disease in sunflower?

Ans. Downy mildew (*Plasmopara halstedii*)

Q.32 Damping-off disease in solanaceous vegetables in nursery is caused by the seed borne nature of ______ pathogen

Ans. *Pythium aphanidermatum*

Q.33 The designated disease, brinjal fruit rot / blight is caused by the organism called ________

Ans. *Phomopsis verans*

Q.34 'Buck eye rot' disease in tomato is caused by the incidence of

Ans. *Phytophthora parasitica*

Q.35 What is the name of the early blight disease causing pathogen in tomato?

Ans. *Alternaria solani* (designated disease)

Q.36 Tomato late blight disease is caused by ________ pathogen

Ans. *Phytophthora infestans*

Q.37 What is the causal organism of tomato leaf spot?

Ans. *Stemphylium solani* (designated disease)

Q.38 What is the name of the viral designated disease in tomato?

Ans. Tobacco Mosaic Virus (TMV) (*Nicotiana virus* 1)

Q.39 What is the disease causing organism for anthracnose or fruit rot in chilli?

Ans. *Colletotrichum capsici* (designated disease)

Q.40 Anthracnose leaf spot in cucurbits is caused by ______

Ans. *Colletotrichum lagenarium*

Q.41 In crucifers, the gray and black leaf spot is caused by the seed borne pathogen ________

Ans. *Alternaria brassicae*

Q.42 What is the seed infection percentage of black rot that causes epidemic in cabbage?

Ans. 0.02%

Q.43 Black root rot in carrot is caused by the pathogen called ________

Ans. *Alternaria radicina*

Q.44 Damping off disease in onion is caused by ________

Ans. *Botrytis allii*

Q.45 Onion purple blotch disease caused by the organism ________

Ans. *Alternaria porri*

Q.46 What is the black rot disease causing organism in crucifers?

Ans. *Xanthomonas campestris pv. campestris*

Q.47 What are all the designated diseases in crucifers like cabbage, cauliflower and Chinese cabbage?

Ans. Black leg - *Leptosphaeria macutans*
Black rot - *Xanthomonas campestris pv. campestris*
Soft rot - *Erwinia caroovora*

Q.48 Bacterial blight in beans is caused by ________

Ans. *Xanthomonas phaseoli*

Q.49 Name the nematode spread through seeds?

Ans. *Anguina tritici* in wheat

Q.50 Indicate the viral diseases spread by seed?

Ans. Bean Common Mosaic Virus (BCMU)
Tomato Spotted Wilt Virus (TSWV)
Cucumber Mosaic Virus (CMV)

Q.51 The methods used to detect the seed borne virus are ________

Ans. Enzyme-Linked Immunosorbent Assay (ELISA)
Serologically Specific Electron Microscopy (SSEM)

Q.52 What is the name of the certificate that a seed should have during import?

Ans. Phytosanitary certificate

Q.53 Who is the father of seed pathology?

Ans. Paul Neergaard

Q.54 Who coined the term 'seed pathology'?

Ans. Paul Neergaard and Mary Noble

Q.55 What is the name of the book written by Paul Neergaard?

Ans. Seed pathology (two volumes)

Q.56 What are the two types of insect feeders in seeds?

Ans. Internal feeders
External feeders

Q.57 Insects that feed on seed and flour materials are termed as ____________

Ans. External feeders

Q.58 Insects that feed on the internal content of seed including embryo are called as ________

Ans. Internal feeders

Q.59 Which type of insects that cause more damage to seeds?

Ans. Internal feeders

Q.60 Give examples for internal feeders in seeds?

Ans. Pulse beetle (*Callosobruchus chinensis* and C. *maculatus*)
Drug store beetle (*Stegobium paniceum*)
Cigarette / tobacco beetle (*Lasioderma serricorne*)
Lesser grain borer (*Rhyzopertha dominica*)
Rice weevil (*Sitophilus oryzae*)
Larger rice weevil / Maize weevil (*Sitophilus zeamais*)
Granary weevil (*Sitophilus granarius*)
Angumois grain moth (*Sitotroga cerealella*)
Tamarind beetle (*Caryedon serratus*)

Q.61 Examples for external feeders are____________

Ans. Red flour beetle (*Tribolium castaneum, T. confusum*)
Khapra beetle (*Trogoderma granarium*)
Saw toothed grain beetle (*Oryzaephilus surinamensis*)
Indian meal moth (*Plodia interpunctella*)
Rice moth (*Corcyra cephalonica*)
Warehouse moth (*Ephestia cautella*)

Q.62 What is the major seed storage pest for pulses?

Ans. Bruchids (pulse beetle) (*Callosobruchus chinensis* and C. *maculatus*)

Q.63 What is the percentage of loss by bruchids?

Ans. 40-50%

Q.64 Wheat, rice and maize seeds are affected by the internal storage pests like ________

Ans. Lesser grain borer (*Rhyzopertha dominica*)
Rice weevil (*Sitophilus oryzae*)
Larger rice weevil / Maize weevil (*Sitophilus zeamais*)
Granery weevil (*Sitophilus granarius*)
Angumois grain moth (*Sitotroga cerealella*)

Q.65 What are the external storage pests for rice, wheat, maize and oilseeds?

Ans. Khapra beetle (*Trogoderma granarium*)
Saw toothed grain beetle (*Oryzaephilus surinamensis*)
Warehouse moth (*Ephestia cautella)*
Indian meal moth (*Plodia interpunctella*)
Rice moth (*Corcyra cephalonica*)

Q.66 Broken seeds/mechanically damaged seeds are preferred by the storage pest called ________

Ans. Red flour beetle (*Tribolium castaneum*)

Q.67 Groundnut pods are affected by the storage pest called ________

Ans. Tamarind beetle (*Caryedon serratus*)

Q.68 What are all the major storage pests for spice seeds?

Ans. Drug store beetle (*Stegobium paniceum*)
Cigarette / tobacco beetle (*Lasioderma serricorne*)

Q.69 Which fumigant is used for the control of storage pest in godowns?

Ans. Aluminium phosphide @ 7 tablets / tone of seeds

Q.70 What is the seed standard for insect-damage ?

Ans. 1.0% for maize and legumes and 0.5% for other seeds.

□□□

Chapter 10

Organizations and Laws

Q.1 Which year was declared as International Year of Seed by UN?

Ans. 1961

Q.2 Who described seed as the carrier of new technologies?

Ans. Feistritzer, 1975

Q.3 Expand ISTA?

Ans. International Seed Testing Association

Q.4 Where ISTA is located?

Ans. Zurich, Switzerland

Q.5 When was ISTA formed?

Ans. 1924

Q.6 Main aim of ISTA?

Ans. Development of accurate and uniform methods for testing and evaluation of seeds.

Q.7 The book published by ISTA?
Ans. International Rules for Seed Testing

Q.8 Journal published by ISTA?
Ans. Seed Science and Technology

Q.9 Who is the current Secretary General of ISTA?
Ans. Michael Muschick

Q.10 Editor of the Journal, Seed Science and Technology?
Ans. Alison A. Powell

Q.11 Specify the name of the certificate issued by ISTA?
Ans. International Seed Analysis Certificate

Q.12 How many countries are members in ISTA?
Ans. 79

Q.13 How many laboratories are members in ISTA?
Ans. 201

Q.14 How many technical committees are there in ISTA?
Ans. 15

Q.15 How many laboratories are accredited to issue ISTA certificates?
Ans. 120

Q.16 Three types of certificates issued by ISTA are ________
Ans. Orange seed lot certificate
Green seed lot certificate
Blue seed sample certificate

Q.17 The colour of the international seed analysis certificate where the sampling and testing have been carried out in the same country?
Ans. Orange seed lot certificate

Q.18 Indicate the colour of the international seed analysis certificate where the sampling and testing is done at different country

Ans. Green seed lot certificate

Q.19 Official languages of ISTA?

Ans. English, French, German

Q.20 Expand the term 'ISF'?

Ans. International Seed Federation

Q.21 Where is the head quarters of ISF located?

Ans. Nyon, Switzerland

Q.22 What is the mission of ISF?

Ans. Facilitate the international movement of seed

Q.23 Expand the term AOSA?

Ans. Association of Official Seed Analysts

Q.24 Where is the AOSA located?

Ans. Ithaca, New York

Q.25 When AOSA was started?

Ans. 1908

Q.26 What is the name of the journal published from AOSA?

Ans. Seed Technology Journal

Q.27 Expand the abbreviation 'UPOV'.

Ans. International Union for the Protection of New Varieties of Plants

Q.28 OECD is called as ________

Ans. Organization for Economic Cooperation and Development

Q.29 Where the head quarters of OECD is located?

Ans. Paris (France)

Q.30 AOSCA is expanded as?

Ans. Association of Official Seed Certifying Agency

Q.31 When AOSCA was established?

Ans. 1919

Q.32 Head quarters of AOSCA is at ________?

Ans. Moline, Illinois (US)

Q.33 Main objective of AOSCA is ______

Ans. Establishment of minimum standards for certification

Q.34 IPPC is expanded as__________

Ans. International Plant Protection Convention

Q.35 IPPC was formed during the year________

Ans. 1952

Q.36 What is the name of the organization available for the international plant health?

Ans. IPPC

Q.37 Where is the head quarter of IPPC located ?

Ans. FAO in Rome

Q.38 FAO is expanded as__________

Ans. Food and Agricultural Organization

Q.39 The Indian institute that tests the imported germplasm seed is ____________

Ans. National Bureau of Plant Genetic Resources (NBPGR), New Delhi

Q.40 NBPGR is located at _________

Ans. New Delhi

Q.41 __________is the institute in India in which the germplasm seeds are stored in seed bank.

Ans. National Bureau of Plant Genetic Resources (NBPGR)

Q.42 The institute which is responsible for worldwide germplasm conservation is ____________

Ans. IPGRI

Q.43 IPGRI is expanded as_________

Ans. International Plant Genetic Resources Institute

Q.44 Where IPGRI is located?

Ans. Rome, Italy

Q.45 The present name of IPGRI is __________

Ans. Biodiversity International

Q.46 When was the name of IPGRI changed ?

Ans. 1st December 2006

Q.47 Some other institutes responsible for germplasm conservation are ________

Ans. International Centre for Tropical Agriculture (CIAT), Colombia

International Institute of Tropical Agriculture (IITA), Nigeria

International Centre for Potato (CIP), Peru

Asian Vegetable Research and Development Centre (AVRDC), Taiwan

Q.48 What is meant by 'CBD'?

Ans. Convention on Biological Diversity

Q.49 'WIPO' stands for _________

Ans. World Intellectual Property Organization

Q.50 'World Intellectual Property Organization" head quarter is located at______?

Ans. Geneva (Switzerland)

Q.51 What is meant by CSTL?
Ans. Central Seed Testing Laboratory

Q.52 Where the CSTL is located?
Ans. Varanasi

Q.53 When the first CSTL was established?
Ans. 1961

Q.54 First CSTL functioned at ___________?
Ans. Indian Agricultural Research Institute (IARI), New Delhi.

Q.55 When was the first CSTL notified?
Ans. 1969

Q.56 NSRTC is the abbreviation of ___________?
Ans. National Seed Research and Training Centre

Q.57 Where is NSRTC located?
Ans. Varanasi

Q.58 When was NSRTC established?
Ans. 2005

Q.59 The laboratory functioning under NSRTC is ___________?
Ans. Central Seed Testing Laboratory (CSTL)

Q.60 When was the CSTL established at NSRTC?
Ans. 2006

Q.61 What is meant by ISST?
Ans. Indian Society for Seed Technology

Q.62 The official journal published from ISST?
Ans. Seed Research

Q.63 The foundation year of ISST?

Ans. 1971

Q.64 The periodicity of the Seed Research journal is __________?

Ans. Half yearly.

Q.65 The number of issues published in a year in Seed Science and Technology journal is __________

Ans. Four (quarterly)

Q.66 'Seed Science Research' journal is published as ______

Ans. Quarterly

Q.67 Who is the publisher of 'Seed Science Research' journal?

Ans. Cambridge University Press.

Q.68 Editor of the journal 'Seed Science Research' is__________

Ans. Professor M. Black

Q.69 What is meant by APSA?

Ans. Asia and Pacific Seed Association

Q.70 APSA is located at _________?

Ans. Bangkok (Thailand)

Q.71 What is the name of the journal / magazine published by APSA?

Ans. Asian Seed and Planting Material (ASPM)

Q.72 What is meant by 'ISSS'?

Ans. International Society for Seed Science

Q.73 What is the official journal of ISSS?

Ans. Seed Science Research

Q.74 Where is the secretariat of ISSS located?

Ans. West Sussex (United Kingdom)

Q.75 Name of the International Organization for Forest Seed Research is _______

Ans. International Union of Forestry Research Organization (IUFRO)

Q.76 Where the secretariat of IUFRO is located?

Ans. Vienna (Austria)

Q.77 The research articles related to seed published in journals are indexed in ______Abstracts?

Ans. Seed.

Q.78 IPR is expanded as __________

Ans. Intellectual Property Right's

Q.79 TRIPS is expanded as ________

Ans. Trade Related Intellectual Property Rights

Q.80 Expand the term GATT?

Ans. General Agreement on Tariff and Trade

Q.81 Seed Act was passed during the year ______

Ans. 1966

Q.82 An act to provide for regulating the quality of certain seeds was called as___

Ans. The Seeds Act, 1966

Q.83 What is known as the 'sanctioning legislation' ?

Ans. The Seeds Act 1966

Q.84 When were the 'Seed Rules' passed?

Ans. 1968

Q.85 What is known as the 'governing legislation' ?

Ans. The Seeds Rules, 1968

Q.86 When were Seeds Act and Seeds Rules implemented in India?

Ans. 1.10.1969

Q.87 The Seeds Act, 1966 consists of ___________sections.

Ans. 25

Q.88 Notification of kinds and varieties for seed certification is done as per section ____of the Seeds Act, 1966

Ans. 5

Q.89 What is the section that provides for establishment of Central Seed Laboratory under Seeds Act?

Ans. Section 4

Q.90 What is the name of person appointed under Section 12 of Seeds Act?

Ans. Seed Analyst

Q.91 Name the section that sanctions for the establishment of Central Seeds Committee?

Ans. Section 3

Q.92 Who is appointed as per section 13 of Seeds Act?

Ans. Seed Inspector

Q.93 A laboratory notified under Subsection 2 of Section 4 of Seeds Act is known as _______

Ans. State Seed Laboratory

Q.94 What is the term period for members of Central Seed Committee?

Ans. Two years

Q.95 What is the section that authorizes the minimum limits of germination and purity as per Seeds Act?

Ans. Section 6

Q.96 What is the section that sanctions the formation of seed certification agency?

Ans. Section 8

Q.97 What is the section that provides for formation of appellate authority?

Ans. Section 11

Q.98 How many times was the Seeds Act amended?

Ans. Once

Q.99 Seeds Act was amended during the year _____,

Ans. 1972

Q.100 Name the act that sanctions for the implementation of Central Seed Certification Board?

Ans. The Seeds (Amendment) Act, 1972

Q.101 How many times was the Seeds Rules amended?

Ans. Three times

Q.102 Seeds Rules were amended during the years _________, ___________,______

Ans. 1973, 1974, 1981

Q.103 Which part of Seeds Rules deals with the function of seed certification agency ?

Ans. Part IV

Q.104 Amendments to Seed Rules were made during ________

Ans. 1981

Q.105 When was 'Seeds Control Order' implemented ?

Ans. 1983

Q.106 Name the legislation under which seeds were declared as Essential Commodity

Ans. Seeds (Control) Order, 1983

Q.107 What was the legislation brought by GOI under the powers of Sec. 2 of Essential Commodities Act?

Ans. Seeds Control Order

Q.108 Essential Commodities Act was passed during which year?

Ans. 1955

Q.109 What was the act brought out by parliament on 30.12.1983?

Ans. Seeds Control Order

Q.110 The main mandate of Seed Control Order is ___________

Ans. Compulsory licensing

Q.111 What is the act which declares that seed dealer must posses a license to sell seeds?

Ans. Seeds Control Order, 1983

Q.112 What is the license fee prescribed for selling seeds as per Seeds Control Order?

Ans. Rs 25

Q.113 Name the form for applying seed dealer license under Seeds Control Order

Ans. Form A

Q.114 As per Seeds Control Order, the seed dealer license is provided under which form

Ans. Form B

Q.115 What is the validity period for seed dealer's license?

Ans. 3 years

Q.116 Is it true that a person convicted under Essential Commodities Act for 3 years can be provided with seed dealer's license?

Ans. Not true.

Q.117 Who is the licensing authority as per Seeds Control Order?

Ans. ADA Seed Inspection

Q.118 What is the maximum time limit for assessing the sample provided by Seed inspector?

Ans. 30 days

Q.119 What is the name of sample sent by seed inspector to STL ?

Ans. Official sample

Q.120 Expand the EPA?

Ans. Environment Protection Act

Q.121 In which year Environment Protection Act was enacted ?

Ans. 1986

Q.122 NPSD is expanded as ________

Ans. New Policy on Seed Development

Q.123 NPSD was enacted during?

Ans. 1988

Q.124 Import and Export of seed is liberalized under __________

Ans. New Policy on Seed Development, 1988

Q.125 What is the law that permitted the import of vegetables and ornamentals under Open General License?

Ans. New Policy on Seed Development, 1988.

Q.126 What is the land frontier that allows for import of seed from Pakistan, Afganistan etc.?

Ans. Wagah border

Q.127 When was the Plant, Fruits and Seeds (Regulation of Import into India) Order passed?

Ans. 1989

Q.128 PPV and FR?

Ans. Protection of Plant Varieties and Farmers Rights Act.

Q.129 In which year PPV & FR Act was passed?

Ans. 2001

Q.130 Where is the PPV & FR Act authority is located?

Ans. New Delhi

Q.131 What is the duration of protection of a variety under PPV and FR Act?

Ans. Trees and vines – 18 years
Others - 15 years

Q.132 What is the nomenclature provided under PPV and FR Act for a variety already notified under Seeds Act ?

Ans. Extant variety

Q.133 BDA is expanded as __________

Ans. Biological Diversity Act

Q.134 Biological Diversity Act Passed during the year _________

Ans. 2002

Q.135 How many agro-climatic zones have been identified as per National Seeds Policy, 2002?

Ans. Five

Q.136 New Seed Bill was proposed during the year ________

Ans. 2004

Q.137 What year "Plants, Fruits and Seeds Order" was enacted?

Ans. 1985

Q.138 What is the other name of "Plants Fruits and Seeds Order" 1985 ?

Ans. Destructive Insect and Pests (DIP) Act.

Q.139 The main objective of DIP Act is ____________?

Ans. To check the seed or plant consignment for its valid import permit.

Q.140 'Royal Commission on Agriculture' was set up in the year____________

Ans. 1925

Q.141 "All India Seed Growers, Merchants and Nurseryman's Association" was established during the year________

Ans. 1946

Q.142 'Grow More Food Campaign (Grow More Food Enquiry Committee)' was launched during the year__________

Ans. 1952

Q.143 The 'Seed Multiplication Team Review' was formed during __________

Ans. 1961

Q.144 The first sorghum hybrid was released in the year_______

Ans. 1964

Q.145 The first bajra hybrid was released during the year_______

Ans. 1965

Q.146 'High Yielding Varieties Programme (HYVP)' was launched during the year___

Ans. 1966

Q.147 'Seed Review Team (SRT)' was set up during the year ___________

Ans. 1968

Q.148 Expand the term TDC:___________

Ans. Tarai Development Corporation

Q.149 TDC was started during the year _______

Ans. 1969

Q.150 The 'Tarai Development Corporation' was later christened as ________

Ans. UP Seeds and Tarai Development Corporation Ltd (1.7.1978)

Q.151 The main motto of TDC is ____________

Ans. Compact area approach for seed production and marketing

Q.152 National Seed Programme (NSP) was initiated during the year __________

Ans. 1974

Q.153 When was the National Seed Project Phase I executed ?

Ans. 1977-78

Q.154 What is meant by NSC?

Ans. National Seeds Corporation

Q.155 Central Seeds Corporation was later named as _______

Ans. National Seeds Corporation

Q.156 Where is the NSC headquarters located?

Ans. New Delhi

Q.157 When has the first public sector corporation, NSC established?

Ans. 1963

Q.158 SFCI is expanded as _________

Ans. State Farm Corporation of India

Q.159 The headquarter of SFCI is located at_________

Ans. New Delhi

Q.160 SFCI was established during the year _________

Ans. 1969

Q.161 When was the first seed testing laboratory in India established?

Ans. 1961

Q.162 When was seed bank scheme launched ?

Ans. 2000

Q.163 Who has the right to send seeds to referral lab?

Ans. Judge in whose court seed judicial proceedings have been filed

Q.164 Where is the 'Directorate of Seeds located' ?

Ans. Mau, UP

Q.165 In India, first seed testing laboratory was established at _________

Ans. IARI, New Delhi.

Q.166 How many seed testing laboratories are there in India?

Ans. 108.

Q.167 How many Indian Seed Testing Laboratories are members of ISTA?

Ans. 21 (7 in public and 14 in private sector)

Q.168 Mention the numbers of ISTA accredited seed testing laboratories in India.

Ans. 6 (All in private sector)

Q.169 Which is the Referral Laboratory for Seed Testing?

Ans. Central Seed Testing Laboratory, Varanasi.

Q.170 Which is the Referral laboratory for testing Bt cotton seeds?

Ans. Central Institute for Cotton Research (CICR), Nagpur.

Q.171 What is meant by reference test?

Ans. Additional seed quality test conducted in another laboratory

Q.172 The "Seed Association of India" is located at __________

Ans. New Delhi

Q.173 "Association of Seed Industries" is located at __________

Ans. Mumbai

Q.174 'Indian Minimum Seed Certification Standards' were established during the year ____

Ans. 1971

Q.175 SSC is termed as__________?

Ans. State Seed Corporation

Q.176 How many SSC's are there in India?

Ans. 15

Q.177 SSCA is expanded as __________

Ans. State Seed Certification Agencies

Q.178 In India, ____________ SSCA's are involved in the regulation of quality seed.

Ans. 20

Q.179 WTO means?

Ans. World Trade Organization

Q.180 India's share in world seed trade is __________

Ans. 3%

Q.181 India occupies the ___________rank in world seed market

Ans. 8th

Q.182 No written contract, but guarantees that farmers will receive a profit i.e. seed will be purchased above the market value of the grain is called ________

Ans. *Laissez faire* approach

Q.183 Compact Area Approach for seed production is synonymous with ________

Ans. Seed Village Concept

Q.184 The most successful location by compact area approach is ________

Ans. Rannebennur, Karnataka

Q.185 ___________is called as 'Seed Valley of India'

Ans. Hyderabad

❑❑❑

Chapter 11

Books on Seed Science and Technology

Q.1 Who gave the book 'Hand buck der semenkunde', the first treaste on seed technological science ?

Ans. Dr.Fedrich Nobbe

Q.2 'Seed Biology' (three parts) book is written by the author______

Ans. T.T.Kozlowski

Q.3 'Viability of Seeds' book is written by ________

Ans. E.H. Roberts

Q.4 Authors of 'Principles of Seed Science and Technology' are ____________

Ans. Lawrence O. Copeland and M. B. McDonald

Q.5 The book on 'Seeds: Physiology of Development and Germination' is written by _________

Ans. J.Derek Bewley and Michael Black

Q.6 'Seed Development and Germination' is written by _______

Ans. Jaime Kigel and Gad Galili

Q.7 'Handbook of Seed Physiology: Applications to Agriculture' is authored by ___

Ans. Roberto L. Benech-Arnold and Rodolfo A. Sanchez

Q.8 'Seed Development, Dormancy and Germination' book written by the authors___

Ans. Kent J. Bradford and Hiroyuki Nonogaki

Q.9 Who authored the book called 'Seeds: Ecology, Biogeography, and Evolution of Dormancy and Germination'?

Ans. Carol C. Baskin and Jerry M. Baskin

Q.10 'The Encyclopedia of Seeds' was written by ___________

Ans. J. Derek Bewley, Michael Black and Peter Halmer

Q.11 'The Ecology of Seeds' book was written by _____

Ans. Michael Fenner and Ken Thompson

Q.12 'Synseeds: Applications of Synthetic Seeds to Crop Improvement' book was edited by ______

Ans. Keith Redenbaugh

Q.13 'Seeds: The Ecology of Regeneration in Plant Communities' is authored by____

Ans. Michael Fenner

Q.14 'Ecology of Soil Seed Banks' is written by the authors____

Ans. Mary Allessio Leck, V. Thomas Parker and Robert Lee Simpson

Q.15 Who is the author of the book 'The Physiology and Biochemistry of Seed Development, Dormancy and Germination'?

Ans. Anwar Ahmad Khan (A.A.Khan)

Q.16 'Seed Science and Technology Laboratory Manual' is written by ______

Ans. Miller B. McDonald

Q.17 Name the author of the book called 'Seed Production: Principles and Practices'?

Ans. Miller B. McDonald

Q.18 'Sprouting Seed Science Projects' is written by the author_______

Ans. Benbow Ann

Q.19 'A Reference Book on Seed Science' is written by ___

Ans. V. Brouver and A. Shtelin

Q.20 Who wrote the book named 'An Introduction of Seed Technology'?

Ans. J.R. Thomson

Q.21 'Seeds: Biology, Development and Ecology' is authored by __________

Ans. S.W.Adkins, S. Ashmore and S.C.Navie

Q.22 'Successful Seed Programs: A Planning and Management Guide' book is written by ______

Ans. Johnson E. Douglas

Q.23 'Seed Planning and Policy for Agricultural Production' is written by _________

Ans. A. Fenwick Kelly

Q.24 'Seed Identification Manual' book is written by ______

Ans. Alexander C. Martin and William D. Barkley

Q.25 'Vegetable Seed Production' book is authored by________

Ans. Raymond A.T. George

Q.26 The book called 'Seed Dormancy in Grasses' is written by the scientist________

Ans. G.M. Simpson

Q.27 The book on 'Seed and their Uses' is written by the authors ________

Ans. Carol Duffus and Colin Slaughter

Q.28 The book on 'Seed Proteins – Biochemistry, Genetics and Nutritive Value' is written by ________

Ans. Werner Gottschalk and Hermann P Muller

Q.29 A book on 'Seed Marketing' was written by the authors________

Ans. Alvin G. Law, Billy R. Gregg, Paul B. Young and Pakkali R. Chetty

Q.30 'Handbook of Seed Physiology' book edited by ________

Ans. R.L. Benech-Arnold and R.A.Sanchez

Q.31 A book on 'Seed Technology and Its Biological Basis' written by________

Ans. M. Black and J.D. Bewley

Q.32 The popular book on 'Seed Technology' was written by the author________

Ans. R.L Agrawal

Q.33 Who wrote the book on 'Principles of Seed Technology'?

Ans. P. K. Agrawal

Q.34 'Handbook on Seed Testing' book was edited by the author______

Ans. P.K. Agrawal

Q.35 'Indian Minimum Seed Certification Standards' book was prepared by the authors________

Ans. N.S. Tunwar and S.V. Singh

Q.36 'Seed Health' is authored by ________

Ans. V. K. Agarwal

Q.37 'Techniques in Seed Science and Technology' was authored by ________

Ans. P.K. Agrawal and M. Dadlani

Q.38 'Seed Storage of Horticultural Crops' book was written by ____________

Ans. S. D. Doijode

Q.39 'Seed Quality: Basic Mechanisms and Agricultural Implications' was written by________

Ans. Amarjit S. Basra

Q.40 The book on 'Handbook of Seed Science and Technology' was authored by ___

Ans. Amarjit S. Basra

Q.41 'Identification of Crop and Weed Seeds' was written by the author________

Ans. A.F. Musil

Q.42 'A Handbook of Seed Testing' was written by ________

Ans. J. Renugadevi, P. Srimathi, R. R. Renganayaki and V. Manonmani

Q.43 Who authored the book on 'Experimental Seed Science and Technology'?

Ans. R. Umarani, R. Jerlin, P. Masilamani, N. Natarajan and A.S. Ponnuswamy

Q.44 'Advances in Seed Science and Technology Volume - 1: Recent Trends in Seed Technology and Management' edited by the authors________

Ans. K. Vanangamudi, N. Natarajan, A. Bharathi, R. Umarani, K. Natarajan and T. Saravanan

Q.45 'Advances in Seed Science and Technology Volume - 2: Quality Seed Production in Vegetables' was edited by ______

Ans. K. Vanangamudi, N. Natarajan, P. Srimathi, K. Natarajan, T. Saravanan, M. Bhaskaran, A. Bharathi, P. Natesan and K. Malarkodi

Q.46 'Advances in Seed Science and Technology Volume - 3: Forest Tree Seed Technology and Management' was edited by______

Ans. K. Vanangamudi, K. Natarajan, T. Saravanan, N. Natarajan, R. Umarani, A. Bharathi and P. Srimathi

Q.47 'Advances in Seed Science and Technology Volume - 4: Forest Tree Seed Production' edited by ______

Ans. K. Vanangamudi, K. Natarajan, T. Saravanan, N. Natarajan, R. Umarani, A. Bharathi and P. Srimathi

Q.48 'Advances in Seed Science and Technology Volume - 5: Quality Seed Production in Spices, Plantation, Medicinal and Aromatic Crops' edited by ___

Ans. K. Vanangamudi and K. Natarajan

Q.49 'Seeds Handbook: Biology, Production, Processing and Storage' is written by the authors ______

Ans. B. B. Desai, P. M. Kotecha and D. K. Salunkhe

Q.50 'Agro's Dictionary of Seed Science and Technology' was written by ______

Ans. Kins Varghese and Gautam Ghosh

Q.51 'Seed Production of Commercial Vegetables' book was written by _____

Ans. Y. S. Sreenivas

Q.52 'Seed Production of Field Crops' was written by _____

Ans. Brajesh Tiwari

Q.53 'Seed Science and Technology' book was written by __________

Ans. Bhanu, T.K. Chandra and Bhatnagar Vasudev

Q.54 Who wrote the another book on 'Seed Science and Technology'?

Ans. A. K. Joshi and B.D. Singh

Q.55 Who is the author of the book 'Seed Science and Technology – Indian Forestry Species ?

Ans. P.E.Bedell

Q.57 'Seed Legislation in India' was compiled by ________

Ans. K. Ramamoorthy, K. Sivasubramaniam and A. Kannan

Q.58 'Seed Technology Readyreckoner' was authored by ________

Ans. K. Ramamoorthy and K. Sivasubramaniam

Q.59 'Recalcitrant Seeds: Causes and Effects' written by the authors_________

Ans. K. Sivasubramaniam, K. Raja and R. Geetha

Q.60 'Stored Grain Pests and their Management' book was written by _________

Ans. B.P. Khare

Q.61 'A Guide to Seed Legislation' was compiled by the authors __________

Ans. C. Ramalingam, K. Sivasubramaniam and A. Vijayakumar

Q.62 'Compendium on Seed Science and Technology' book was edited by ________

Ans. V. Krishnasamy, P. Balamurugan, N. Natarajan, A.S.Ponnuswamy, P. Srimathi and T.S. Raveendran

Q.63 'A Text Book on Seed Pathology' was written by ________

Ans. D.K. Jha

Q.64 'Principles of Seed Certification and Testing' written by the author ________

Ans. N.P. Nema

Q.65 'Fundamentals of Plant Breeding and Hybrid Seed Production' was authored by __

Ans. Rattan Lal Agrawal

Q.66 'Seed Science and Technology' book was written by the authors______

Ans. Subir Sen and Nabinananda Ghosh

Q.67 Book on 'Seed Science and Technology' was written by _____

Ans. Gurnam Singh

Q.68 'Synthetic Seeds for Commercial Crop Production' book was written by ______

Ans. U. Kumar

Q.69 'Vegetable Seed Production Principles' was written by______

Ans. Prem Singh Arya

Q.70 'Seed Health Testing - Principles and Protocols' book was authored by _______

Ans. Karuna Vishunavat

Q.71 'Research Perspective in Seed Production and Storage' book is written by ___

Ans. G.N. Kulkarni

Q.72 'Seed Technology and Seed Pathology' was authored by __________

Ans. Y.S. Sreenivas

Q.73 Book on 'Seed Technology and Seed Pathology' was authored by________

Ans. Uma Shankar Singh

Q.74 "Seeds: their Conservation Principles and Practices" was written by ______

Ans. Veena Gupta

Q.75 'A Dictionary of Seed Technological Terms' was compiled by ______

Ans. C. Dharmalingam, K. Sivasubramaniam and Shiv K. Yadav

Q.76 'Seed Treatment' was written by ______

Ans. K.A. Jeffs

Q.77 'Seed Physiology' book was edited by ______

Ans. K. Vanangamudi, Mallika Vanangamudi, M. Djanaguiraman, K. Natarajan, K. Raja, N. Natarajan and T. Saravanan

Q.78 'Hybrid Seed Production in Field Crops' was written by ______

Ans. N.C.Singhal

Q.79 'Seed Science and Technology' book was written by the author______

Ans. N.C.Singhal

Q.80 'Seed: A Global Perspective' book is written by ______

Ans. G. Kalloo, S.K. Jain, Alice K. Vari and Umesh Srivastava

Q.81 'Seed Industry in India' book was written by the authors ______

Ans. Gurdev Singh, S.R. Asokan and V.N. Asopa

Q.82 'Synthetic Seed: A Challenging Technology in Plant Propagation, Transportation and Conservation' book was written by ______

Ans. Bidhan Roy

Q.83 'Somatic Embryogenesis and Synthetic Seed' book was written by ______

Ans. Y. P. S. Bajaj

Q.84 'X-Radiography of Seeds' book written by the authors ________

Ans. R. Umarani, A. Venkatesh and K. Vanangamudi

Q.85 Who edited the book on 'Recent Techniques and Participatory Approaches on Quality Seed Production'?

Ans. K. Vanangamudi, N. Natarajan, A. Bharathi, P. Natesan, R. Jerlin, K. Natarajan and T. Saravanan

Q.86 'A guide to Forest Seed Handling – with special reference to the Tropics' book compiled by the author________

Ans. R.L.Willan

Q.87 'Plant Breeding' book is written by the author________

Ans. B.D. Singh

Q.88 Who is the author of 'Fundamentals of Seed Pathology'?

Ans. Karuna Vishunavat

Q.89 'Vegetable Seed Production and Hybrid Technology' book was written by______

Ans. P. Hazra and M.G. Som

Q.90 'Vegetable Crops-Breeding and Seed Production' was written by ______

Ans. M.S. Fageria, P.S.Arya and A.K. Choudhary

Q.91 A book on 'Principles of Seed Technology' was written by ______

Ans. G.M. Kulkarni

Q.92 'Hybrid Seed Production' book was edited by ________

Ans. H.C. Singal

□□□

About the Authors

Dr. K. Raja, Assistant Professor, Seed Science and Technology is an avid reader and budding scientist. He has ten years experience in research, teaching, extension and handled several courses on seed science and technology. He did his doctorate in recalcitrant seed physiology of arecanut seed which forms genesis of the publication of a book "Recalcitrant Seed". He has published several international and national articles and currently he is researching the use of cashew nut shell liquid for controlling bruchids in the Department of Seed Science and Technology, Agricultural College and Research Institute, Madurai, Tamil Nadu.

Dr. K. Sivasubramaniam is scientist of many hues. He did his graduate, post graduate and doctorate in the prestigious Tamil Nadu Agricultural University, Coimbatore. With twenty two years spent in seed production, teaching, and extension activity. He is well known in the field of seed science and has authored five books that have been well received by the peer community. He has also to his credit several schemes and research projects. Presently he is the Professor and Head of Department of Seed Science and Technology, Agricultural College and Research Institute, Madurai, Tamil Nadu.

Dr. R. Geetha is Associate Professor in the Department of Seed Science and Technology, Agricultural College and Research Institute, Madurai, Tamil Nadu. She has a wide range of experience spanning over 15 years. Her knowledge spreads over agricultural and horticultural crops. She has published several scientific articles and manuals for undergraduate courses. Her primary focus is on seed priming techniques.